ROUTLEDGE LIBRARY EDITIONS:
NUCLEAR SECURITY

Volume 28

THE LOGIC OF NUCLEAR TERROR

THE LOGIC OF NUCLEAR TERROR

Edited by
ROMAN KOLKOWICZ

Routledge
Taylor & Francis Group
LONDON AND NEW YORK

First published in 1987 by Allen & Unwin, Inc.

This edition first published in 2021
by Routledge
2 Park Square, Milton Park, Abingdon, Oxon OX14 4RN

and by Routledge
52 Vanderbilt Avenue, New York, NY 10017

Routledge is an imprint of the Taylor & Francis Group, an informa business

British Library Cataloguing in Publication Data
A catalogue record for this book is available from the British Library

ISBN: 978-0-367-50682-7 (Set)
ISBN: 978-1-00-309763-1 (Set) (ebk)
ISBN: 978-0-367-53245-1 (Volume 28) (hbk)
ISBN: 978-1-00-308111-1 (Volume 28) (ebk)

ROMAN KOLKOWICZ, Editor

The Logic of Nuclear Terror

Published under the auspices of
The University of California Project on Politics and War

Boston
ALLEN & UNWIN
London Wellington Sydney

Allen & Unwin, Inc.
8 Winchester Place, Winchester, MA 01890, USA

The U.S. Company of
Unwin Hyman Ltd,
P.O. Box 18, Park Lane, Hemel Hempstead, Herts HP2 4TE, UK
40 Museum Street, London WC1A 1LU, UK
37/39 Queen Elizabeth Street, London, SE1 2QB, UK

Allen & Unwin Australia Pty Ltd,
8 Napier Street, North Sydney, NSW 2060, Australia

Allen & Unwin (New Zealand) Ltd, in association with the Port Nicholson Press Ltd
Private Bag, Wellington, New Zealand

Library of Congress Cataloging-in-Publication Data

Kolkowicz, Roman.
The logic of nuclear terror.

Includes bibliographies and index.
1. Nuclear warfare—Congresses. I. Title.
U263.K65 1987 355'.0217 86–22317
ISBN 0–04–497031–5
ISBN 0–04–497032–3 (pbk.)

British Library Cataloguing in Publication Data

The logic of nuclear terror.

1. Nuclear warfare 2. Strategy
I. Kolkowicz, Roman
355.4'307'01 U263

ISBN 0–04–497031–5
ISBN 0–04–497032–3 Pbk

Manufactured in the United States of America

For my daughters
SUSAN, LISA, AND GABRIELLA

CONTENTS

ACKNOWLEDGMENTS

Several foundations and institutions generously supported the conference "The Calculus of Terror: Nuclear Strategy and Its Discontents," from which this volume derives. We are grateful to the Rockefeller Foundation for providing the splendid Villa Serbelloni for the use of the conference; the Alfred Sloan Foundation, the Fritz Thyssen Stiftung, the Carter Center at Emory University, the Institute on Global Conflict and Cooperation of the University of California, and the Center for International and Strategic Affairs at UCLA have generously supported this endeavor. Special recognition is due Donna Beltz, the Conference Coordinator, who skillfully managed conference administration and publications.

In addition to this volume, other papers from the conference will be published as a special issue of the *Journal of Strategic Studies* and in a book by Frank Cass Publishers.

CONTRIBUTORS

Roman Kolkowicz is a Professor of Political Science at the University of California, Los Angeles, where he was the Founding Director of the Center for International and Strategic Affairs, and is currently Director of the Project on Politics and War. He is the author of *The Soviet Military and the Communist Party*. His recent works include *National Security and International Stability* and *Arms Control and International Security*.

Ken Booth is a Professor of International Politics at the University College of Wales, Aberystwyth. He is the author of *Strategy and Ethnocentrism* and co-author of *Contemporary Strategy*.

Lawrence Freedman is a Professor of War Studies at Kings College, London. He is the author of *The Evolution of Nuclear Strategy* and *U.S. Intelligence and the Soviet Strategic Threat*.

Colin S. Gray is President of the National Institute for Public Policy in Washington, D.C. His recent works include *Arms Control and European Security* and *The Nuclear Strategy and National Style*.

Robert Jervis is a Professor of Political Science at Columbia University. He is the author of *The Illogic of American Nuclear Strategy* and *Perception and Misperception in International Politics*.

Benjamin S. Lambeth is a senior staff member at the RAND Corporation and author of numerous monographs and studies on the Soviet military and Soviet foreign and defense policies.

Joseph S. Nye, Jr., is a Professor of Political Science at Harvard University, where he is Director of the Center for Science and International Affairs. His recent works include *The Making of America's Soviet Policy*.

George H. Quester is a Professor of Political Science at the University of Maryland, College Park. He is the author of *American Foreign Policy: The Lost Consensus* and of *Deterrence before Hiroshima*.

Arthur A. Stein is a Professor of Political Science at the University of California, Los Angeles. He is the author of *The Nation at War* as well as numerous articles on questions of International Relations and United States Policy.

Introduction

ROMAN KOLKOWICZ

The United States and the Soviet Union presently possess the equivalent of 8 billion tons of TNT in their strategic arsenals. This vast destructive potential, shared in approximate balance by the two superpowers, is the equivalent of 640,000 times the explosive power of the Hiroshima atomic bomb. It has been estimated that 400 megatons (a mere 5 percent of the total arsenal) dropped on urban and industrial targets in the Soviet Union would destroy two-thirds of the USSR's urban population and three-quarters of its industry; even greater devastation would result if 400 megatons were dropped on the United States.[1]

Even though it is clear that nuclear weapons have radically challenged traditional approaches to war and strategy, efforts to create an acceptable and viable strategy for their management have been anything but easy and reassuring. During the past three decades, however, following an initial period of adaptation and accommodation to the radical implications of nuclear weapons, a consensus has evolved among Western academic and government experts on a theory for the management of national security, the promotion of international stability, and the avoidance of nuclear war. This consensus on strategic deterrence policy has centered since the 1960s on the doctrine of mutual assured destruction (MAD). This doctrine is characterized by an emphasis on rational decision making by top national leaders, those entrusted with the management of national security in a very dangerous world. According to this rational, unitary-actor conceptualization of strategic decision making, nuclear capabilities have vital deterring, dissuasive, and diplomatic–political utilities; their actual military employment is considered irrational and highly undesirable under the assumption that each side would suffer certain and devastating destruction in the event of either a first or even a retaliatory nuclear strike by its roughly equal adversary. Victory (that is, defeat of the adversary) is no longer thought to be a rational objective. According to MAD, nuclear weapons serve a single purpose: to deter an attack by one superpower against the other. Mutual deterrence based on mutual vulnerability has thus become the foundation of modern nuclear strategy.

In recent years, consensus on MAD has been challenged by experts both in and outside the government. Some of these critics argue that a mutual assured destruction doctrine of deterrence is no longer viable; they find its inherent aspects of "consensual vulnerability"[2] to be immoral, since they imply that in the name of strategic stability "killing people is good" (countervalue targeting required by the logic of assured retaliation in MAD), whereas "killing weapons is bad" (counterforce targeting considered to undermine strategic stability and to be indicative of first-strike or preemptive intentions). Others point to the so-called inherent aggressiveness of Soviet leaders and their continued arms buildups, seemingly unconstrained by the imperatives of mutual assured destruction doctrine and by the requirements and logic of deterrence stability. Still others maintain that in order to make United States deterrence policy more credible to the Soviet Union, the United States must move beyond the passive deterrent position of mutual assured destruction—to prepare for war fighting and to consider adopting prevailing and countervailing strategies of protracted nuclear war.

Clearly the golden age of deterrence has passed, that brief period of intellectual ferment and political credibility located during the late 1950s and early 1960s when the promise of harnessing and controlling the nuclear Moloch was high. Through a combination of deterrence, arms control, and crisis management, the advocates of the new science of war had promised to safeguard national security, to provide a stable international environment, and to develop a rational decision-making process for the management of national interests in a hostile nuclear world. "In 1961, the promise was high," observed a noted strategic analyst. "The civilian strategists came to Washington to assume influential roles" in the development of strategic and foreign policies during the Kennedy–McNamara era. By the 1970s, however, such high hopes had been dispelled: "It is fair to say that their performance had not lived up to their promise. And that's putting it mildly."[3]

It is now clear that the erstwhile promises and pretensions of the nuclear whiz-kids and wizards of Armageddon have not been fulfilled. Arms control has neither curbed nor even slowed the vast arms buildup; crisis management has shown itself to be imperfect and fallible; and the centerpiece of this nuclear strategic science, the doctrine of mutual assured destruction, is under assault from all quarters. Critics maintain that whereas the political and technological context of international politics and strategic relations has changed profoundly, our thinking about war and its prevention remains rooted in the past, ossified around a deterrence paradigm that is no longer relevant. Among the reasons cited for this ossification of strategic theory are:

1. The persistent dominance of economic models in the study of strategy. The assumption that international conflict can be analyzed in terms of rational economic–strategic decision making was central to the development of strategic theory, but it is far less useful to theorists who emphasize policy prescription over model building.
2. The transition from strategic "prophet" to "courtier." Virtually "all the basic ideas and philosophies about nuclear weapons and their use have been generated by a few civilians working quite independently of the military."[4] Once these strategists had entered the corridors of power, they ceased to think creatively. "The United States has been living off its strategic theoretical capital ever since."[5]
3. The bureaucratization of strategic science. Attempts to free the development of strategy from the fetters of excessive military–bureaucratic controls and influence, and thus to encourage participation by academics in the development and improvement of strategy, led to the spawning of many think tanks. However, think tanks "as exemplified by the Rand Corporation, began to manifest dual loyalty" and, according to its critics, tended to produce "both irrelevant policy advice and poor scholarship."[6]

With this in mind, distinguished contributors to this volume have endeavored to answer some of the following questions:

1. What is the historical validity, theoretical vitality, and policy relevance of nuclear deterrence theories and doctrines?
2. In what ways have recent technological and policy changes affected the original concepts of nuclear war and deterrence strategy?
3. Are there alternate ways to think about strategy in the contemporary nuclear context that might be more useful?[7]

THE LEGACY OF BERNARD BRODIE AND THE EVOLUTION OF MAD

Bernard Brodie, to whom this volume is dedicated, was a pioneer of modern strategic studies in the nuclear era whose work has powerfully influenced generations of strategists and decision makers. He was the first to perceive and publicly articulate the revolutionary implications of nuclear weapons for war and peace and for the management of international politics in a dangerous world. In 1946, at the very dawn of the nuclear era, he defined the essential paradox and inescapable truth of the nuclear condition: "Thus far, the chief purpose of our military establishment has been to win wars. From now on, its chief purpose must be to avert them. It can have no other useful purpose."[8]

In his lifelong concern for the intellectual integrity and policy relevance of strategic studies in the nuclear era, Brodie insisted that "we need people who will challenge and dissect the prevailing dogmas" of strategic fashions and fads. Above all, reflecting the teachings of Clausewitz whom he greatly admired, Brodie urged the "need to stress the superior importance of the political side of strategy to the simply technical and technological."[9] "The most important single idea in [Clausewitz's] *On War* . . . the one that makes it great, is the idea that war must never be an act of blind violence, but must be dedicated to achieving the supreme goal of statecraft."[10] He considered Clausewitz's assertion that "war has its own language, but not its own logic" to be the "single most important idea in all strategy" because the use of military force must at all times be determined by political decision and rational purpose. And so he emphasized the need to be aware of the "inevitable limitations and imperfections of scientific method in strategic analysis and decisionmaking" and to understand that "the most basic issues of strategy often do not lend themselves to scientific analysis . . . because they are laden with value judgements and therefore tend to escape any kind of disciplined thought." In other words, "war would only be senseless destruction if it were not in pursuit of some valid political objective."[11]

Brodie, like Clausewitz before him, was concerned lest strategy become an abstract, excessively theoretical science of war and thus become detached from its guiding and constraining political context. Above all, Brodie asserted, before leaders consider the use of military force, they must ask: "De quoi s'agit-il?" Can nuclear weapons still be considered militarily usable and politically rational? "Should we abandon deterrence strategies in favor of war-winning strategies? But what does this mean?"[12]

It should be noted that there was resistance after World War II to Brodie's revolutionary formulation of the nuclear condition and his advocacy of a deterrence-only, nonmilitary use of nuclear weapons. Brodie's views, which rejected the possibility of meaningful victory in a nuclear confrontation once the Soviet Union obtained rough parity with the United States, were later challenged by such influential insiders as William L. Borden and Paul Nitze. Even though these critics conceded that nuclear war would cause enormous devastation, and while they understood the potential deterrence value of these weapons, they nevertheless considered the use of such weapons both conceivable and rational. Borden's and Nitze's positions, which reflected the views of the political and military mainstream in the United States at that time, advocated an American strategy that was similar to that of the prenuclear era: they maintained that victory in a nuclear confrontation was both rational and obtainable.[13]

For Brodie and others, however, the use of nuclear weapons was inconceivable, and thus military victory in nuclear war was not possible. These views were eventually accepted in the 1960s by the official defense community and decision makers and became the foundation for American strategic policies. As Robert McNamara noted,

> Having spent seven years as Secretary of Defense . . . I do not believe we can avoid serious and unacceptable risk of nuclear war until we recognize . . . that nuclear weapons serve no military purpose whatsoever. They are totally useless—except only to deter one's opponent from using them.[14]

By the 1970s, however, noticeable changes emerged among United States officials and experts concerned with strategy and with the credibility of MAD as a viable doctrine for the management of superpower relations. These concerns were motivated by several developments: a steady and worrisome Soviet arms buildup that threatened to exceed parity and the assumed requirements of strategic stability; a perception of renewed Soviet aggression and military pressure in the various parts of the Third World; and a Soviet unwillingness to establish *linkage* between normalization of superpower relations and accommodation on other issues of disagreement. There was a growing perception in the United States that the forces and strategies developed under the assumptions of MAD doctrine were inadequate to protect United States interests and to safeguard its security in the face of a massively expanding Soviet military capability.

Whatever the specific reasons, the official United States departure from the policies of mutual assured destruction was first signalled by then Secretary of Defense James Schlesinger in 1974 and was continued as countervailing strategy under Presidents Carter and Reagan. United States strategy became "heavily oriented to fighting a nuclear war, even a protracted nuclear war," and the list of strike targets was expanded to include not only a minimum of two hundred major Soviet cities, but also Soviet missiles in hardened silos. The Reagan administration moved even further away from MAD doctrine by placing greater emphasis on providing nuclear options during a protracted war. Under these new guidelines, United States nuclear capabilities "must prevail even under the conditions of a prolonged war" and be able to "force the Soviet Union to seek earliest termination of hostilities on terms favorable to the United States." United States official and informed public opinion has thus returned to acceptance of the concept of fighting and winning a nuclear war in the traditional sense of the terms.[15]

What about the Soviet side of this strategic equation? Do Soviet leaders accept and follow the logic and policies of mutual assured destruction, or do they pursue other strategic doctrines and policies? Since the very core of deterrence theory and MAD doctrine rests on mutual perceptions and consequent behavior (and thus on each side's acceptance of the "rules of the game" of deterrence), the question of the Soviet position is critical. Given the protracted antagonism and distrust of the superpowers toward one another, can we safely assume a Soviet willingness to abide by such rules?

During the cold war period, there was little question in the West regarding Soviet aggressive behavior and intent. Belligerent, declaratory Soviet policy and threatening behavior, which culminated in the Cuban missile crisis, did little to change Western views. The strategic community in the West broadly agreed that Soviet strategic doctrine and policy was highly aggressive and offensive, and such agreement included the possibility of a preemptive Soviet nuclear strike under certain conditions. Moreover, many Western strategists and public figures were strongly persuaded that Soviet military and political leaders believed that nuclear war was still fightable and winnable, "if the imperialists initiate it," and that regardless of benign declaratory Soviet policy that supported arms control and detente, Soviet strategic planning was based on these beliefs. As Benjamin Lambeth argues in his chapter in this volume, whereas Soviet declaratory policy has in recent years followed the so-called Tula Line (announced in a speech by Brezhnev in 1977), actual Soviet military planning, weapons acquisition, and strategic doctrine do not reflect a significant change in Soviet behavior. He maintains that the accommodational Tula Line, which disavows Soviet preemption, first use, and the possibility of victory and accepts the principle of strategic equality, is merely a cynical and self-serving Soviet manipulation of Western public opinion. In actuality, he concludes, Soviet strategic policy continues unchanged in the old aggressive, offensive, and preemptive modes. Official United States defense and strategic policies clearly reflect such perceptions of unchanged Soviet strategic planning, as evidenced by the massive arms buildup initiated by the Carter administration and accelerated by Reagan.

ALTERNATIVE PERSPECTIVES

If deterrence and MAD are considered inadequate to contain Soviet aggression and to protect United States national security, how useful and credible are some of the proposals for change? How does one respond to persistent calls for a renaissance in strategic thinking now that the golden age of deterrence is gone and its legacy is considered by many to be largely irrelevant for dealing with the changed strategic–political circumstances of superpower relations?

Clearly, the most dramatic proposals for changing the nuclear deterrence stalemate and going beyond the "immoral" aspects of MAD are contained in the Star Wars proposals of President Reagan. The announced Strategic Defense Initiative (SDI) will turn the United States away from a passive deterrence-only policy (in which the total vulnerability of the United States and Soviet populations ensure their total security) toward an active defense of the country. While SDI has fired imaginations, it has not met with universal approval and acceptance, however. Whereas Reagan administration spokesmen describe SDI as a "more morally and intellectually satisfying" strategy than MAD because the former would "protect the populations rather than avenge them,"[16] others express "serious reservations, to put it mildly, about SDI, on technological, military–strategic and political grounds."[17] Former Secretary of Defense James Schlesinger describes himself as "a skeptic" and remains "highly doubtful about the Strategic Defense Initiative—its pace, its prospective military utility, its costs, and particularly, its more ethereal versions of strategic defense."[18] However, most people, including skeptics of SDI, tend to agree with its advocates on one key point: despite the caveats, "we can't get the SDI genie back into the bottle" and must therefore proceed with our SDI research program, particularly since the Soviets themselves have been spending large amounts of resources on strategic defense research programs. The enormous complications likely to result from partial or full deployment of such technologies in a defense–offense strategic posture are disturbing, and no satisfactory proposals or solutions for overcoming these complications have thus far been provided by SDI advocates (save the less than credible suggestion that the United States share SDI technology with the Soviet Union).

The chapters that follow cluster around four rather well-defined positions concerning the soundness and desirability of present deterrence theory and the mutual assured destruction doctrine:

1. *There is no escape from the decades-old strategy and policy of deterrence based on offense-domination inherent in the mutual assured de-*

struction doctrine. Such technological innovations as SDI are like fool's gold, raising false hopes of stability and survival while in reality introducing dangerous destabilizing factors into superpower strategic and political relations. Deterrence seems to have worked for several decades, and although it is not an ideal means to achieving international stability and national security, it is still the only viable and coherent strategy, given the dynamics and realities of the superpower adversary relationship. Deterrence must therefore be made stronger, more stable, more cooperative, and less threatening.

George Quester and Robert Jervis in particular forcefully support the continued utility of the deterrence paradigm. In considering the policy relevance and theoretical soundness of deterrence and MAD, Jervis identifies some of the dilemmas confronting strategists and statesmen. He distinguishes between good and bad ideas in both the old and the new approaches to nuclear strategy. Among the bad ideas, he cites beliefs that mutual superpower vulnerability can be overcome either by means of such technological fixes as SDI or by proposals to abolish nuclear weapons altogether. He is concerned with the persistent and recently intensified "conventionalization" of nuclear weapons, which makes their use more likely, or at least less forbidding. On the other hand, he applauds some good ideas for strengthening strategic stability among the superpowers through reduction of their sense of vulnerability, specifically policies and weapons that enhance one nation's security without decreasing significantly the security of the other. Here, he addresses the troubling phenomenon of the "security dilemma," whereby one country's attempt to increase its own security is unavoidably perceived by other nations as decreasing theirs. On the whole, he finds technological innovation in the weapons of war (or in the methodological and doctrinal flourishes) to have a very limited impact on the security dilemma, on the requirements of strategic stability, or on the fundamental condition of mutual vulnerability. Instead, he urges the improvement of deterrence stability and credibility by emphasizing not only the retaliatory assured-threat element in superpower strategic relations, but also the credibility of superpower cooperation. In the nuclear era, the only security is mutual security.

2. *Deterrence does not work well; it constrains the United States but not the Soviet Union, which rejects the ideas of self-constraint and self-deterrence implicit in the United States practice of deterrence.* Moreover, the offense-dominated deterrence of MAD is not truly a strategy in the historical sense, because it stops working when it would be most urgently needed, at the outbreak of hostilities. It is also immoral. SDI is the most useful way at present for breaking out of this nuclear

stalemate: it provides for mutual survival rather than mutual insecurity and assured incineration; it also reduces the offensive weapons inventories and emphasizes defense. The United States should therefore proceed with its research and development programs and, at the same time, offer the Soviet Union incentives to negotiate and cooperate; to reduce offensive weapons, vulnerability, and threats; and possibly to deemphasize strategic preemptive doctrines that are highly threatening to mutual stability. The chapters by Colin Gray and Benjamin Lambeth illustrate this position.

3. *Deterrence does not work; it merely legitimates and encourages dogmas and ideologies that in effect demand ever-higher levels of arms development and deployment in the name of vaunted strategic stability.* Deterrence ideologies generate arms races and confrontational moods, without providing additional security or stability. Neither technological fixes nor doctrinal conceptual–methodological "innovation" or "fine tuning" seriously addresses the fundamental problem of nuclear vulnerability and insecurity; the most sensible and rational way out of the present dead end of deterrence is through political and diplomatic means. Ken Booth clearly embraces this interpretation.

4. *Deterrence theories and policies should focus primarily on internal dynamics and context rather than on bipolar strategic and political causative factors from the international environment.* In searching for explanations for the contemporary strategic and political stalemate, several authors emphasize certain factors within the respective superpower systems for a better understanding of what drives these processes. Eschewing the usual explanations of external threat as the central motivator of strategy and politics, Arthur Stein describes SDI as largely motivated by Reagan's domestic politics and partisan concerns. Stein asserts that national security decision making, at least in the pluralistic open societies of the West, is not a depoliticized process left largely to strategic experts. Accordingly, both the very nature of deterrence doctrines and their political implications are contested. Given this reality, politics will determine defense and strategic policies as much as objective analysis and prescriptions of experts.

For example, Stein maintains that the SDI initiative was intended primarily to solve a domestic political problem: it was designed to generate popular and political support for Reagan's massively escalating defense budgets and to help him reposition himself on the nuclear issue. The president, Stein argues, could thus emphasize his passionate commitment to peace and to a world without nuclear weapons, while implying that those who rejected his vision would continue to rely on the "immoral threat of mass murder." This would enable Reagan to avoid serious arms control negotiations with the Soviet Union (negotiations that would undercut political momentum and public support for bigger defense budgets) until his massive rearmament programs were in place. Thus, decisions rooted in domestic or even partisan politics have a powerful impact on the international arena; in the process, they may also set into motion radical changes in widely accepted and fundamental premises and policies.

Kolkowicz considers the main motive force driving United States and Soviet strategic theories and policies to stem less from the real world of superpower confrontations, than from their different historical, political, and sociocultural traditions. In the Soviet case, they are powerfully influenced by their unique war-torn history, traditions of fatalism, and *kto-kovo* distrust of claims to political rationality and benign motives of adversaries in conflict situations.[19]

THE FUTURE

The concluding part of this volume considers the future of deterrence. Joseph Nye, in his chapter on the long-term future of deterrence, observes that most strategists have generally ignored this question and left it largely to antinuclear peace activists, whose wishful or apocalyptic approaches lead to their easy dismissal and thus reinforce the strategists' tendency to avoid the subject. This has led to a theoretical impoverishment of the field and to excessive concentration on rational models of deterrence that tend to be static and ahistorical. Strategists have paid insufficient attention to situations in which rational choice—due to constraints brought about by unanticipated developments and circumstances, or to factors other than those accounted for by rational choice models of deterrence—becomes virtually meaningless.

Nye considers a future in which states and decision makers could reduce their reliance on nuclear weapons and therefore lower the risks of nuclear war. He suggests that the introduction of innovative technologies has helped reduce reliance on tactical nuclear weapons and on "city-busting" strategic weapons, but that technology alone will not achieve the objective of desirable future utopias. The political path must also be taken in order to achieve a stable balance of power: what is needed is a more sophisticated and long-range perception of self-interest, an ability to learn from experience, and a realization that the superpower conflict–cooperation relationship will indeed prevail, at least in the foreseeable future.

Ken Booth examines what he calls the "ideology of deterrence" and its negative connotations and consequences, including justifications and rationalizations for ever-higher levels of arms acquisition and deployment in the name of strategic stability. His essay is in part a kind of meditation on the sorrows of nuclear madness that pervade the world—a system out of control. He vigorously challenges the fundamental assumptions, analyses, and prescriptions of the mainstream and the more extreme advocates of deterrence, and he rejects the credibility of most of their claims to providing a modicum of stability and security in a dangerous nuclear world. He believes that sanity and security lie not in technological or methodological fixes and panaceas, but rather in the political arena.

The vastly differing views of the scholars and experts who have contributed to this volume reflect the uncertainties and absence of a strategic consensus that characterize contemporary thinking about nuclear deterrence theory. In the final analysis, the arguments and debates presented in this volume do not yield any easy or quick solutions to the nuclear dilemma. The authors generally agree that technological fixes or methodological innovations in crisis management are of limited relevance; they also agree that political solutions may lead us, if not into the promised land of eternal peace and absence of war, then at least into peaceful coexistence on this dangerous planet. We may yet be able to agree with Bernard Brodie when he asserted that "it is really marvelous how modern is the ring of what Clausewitz had to say on the subject": politics must ultimately guide and control the arms and strategies of war.[20]

NOTES

1. See Bruce Blair, *Strategic Command and Control: Redefining the Nuclear Threat*, The Brookings Institution, 1985, ch. 2.

2. See Fred C. Ikle, "Nuclear Strategy: Can There Be A Happy Ending?" in *Foreign Affairs*, Spring 1985.

3. Colin Gray, "What RAND Hath Wrought?" *Foreign Policy*, n. 4, Fall 1971.

4. Bernard Brodie, "The Development of Nuclear Strategy" in *National Security and International Stability*, ed. Bernard Brodie, Michael Intriligator, and Roman Kolkowicz (OGH Publishers, 1983), p. 7.

5. Gray, "What RAND Hath Wrought?"

6. Ibid.

7. While the preparations for the Bellagio conference of scholars and specialists were underway, I was invited by former Presidents Jimmy Carter and Gerald Ford to assist them in an analogous enterprise concerned with a reexamination of the relevance of strategic deterrence and arms control policies in the management of superpower relations. Several international conferences, chaired by the two presidents, took place in 1984–85. These

conferences, whose participants were mostly high public officials and political and military leaders from several administrations, as well as from the Soviet Union, China, and other countries, provided an excellent complementary perspective on these complex issues. It is noteworthy that the general conclusions of these conferences concerning the state of superpower political–strategic relations and the roles of existing strategic doctrines were strikingly similar to the conclusions reached by the participants of the Bellagio conference.

See Roman Kolkowicz and Ellen Mickiewicz, ed., *The Soviet Calculus of Nuclear War* (Lexington Books, 1986); and Ellen Mickiewicz and Roman Kolkowicz, *Arms Control and International Security* (Praeger, forthcoming).

8. Bernard Brodie, ed., *The Absolute Weapon* (New York: Harcourt Brace, 1946), p. 76.

9. Bernard Brodie, "Why Were We So (Strategically) Wrong?" *Foreign Policy*, Winter 1971, pp. 151–62.

10. Bernard Brodie, "On Clausewitz: A Passion for War", *World Politics*, January 1973, p. 306.

11. Brodie, "The Development of Nuclear Strategy", p. 12.

12. Ibid., p. 13.

13. See the chapters in this volume by Robert Jervis and George Quester.

14. Cited in *The New York Times*, 15 September 1983, p. A–27.

15. Blair, *Strategic Command and Control.*

16. Secretary of the Navy John Lehman, Jr., cited in Mickiewicz and Kolkowicz, *Arms Control and International Security*, Introduction.

17. Former Secretary of Defense Harold Brown, cited in *Arms Control and International Security*, ibid.

18. Former Secretary of Defense James Schlesinger, cited in *Arms Control and International Security*, ibid.

19. The evident complexity of the strategic nuclear dilemma in a world of massively proliferating weapons and military technologies is sobering. A sense of pessimism at times pervades the chapters in this volume as authors search for ways in which to subordinate proliferating weapons to rational and coherent political and strategic purposes.

This volume thus reflects the pessimism of the deliberations among political and military leaders at the Carter–Ford conferences. Those participants concluded that the record of United States–Soviet arms control negotiations is disappointing; the promise of such technical fixes as SDI is a chimera; and efforts to conciliate United States and Soviet differences on certain issues are too often met with political parochialism and grandstanding, with each side blaming the other for the troubles in the world.

See statements by Ambassador A. Dobrynin, Henry Kissinger, Helmut Sonnenfeldt, and William Hyland, among others, in *Arms Control and International Security*, ibid.

20. Brodie, *The Absolute Weapon.*

PART I

Historical and Theoretical Problems

CHAPTER 1

Intellectuals and the Nuclear Deterrence System

ROMAN KOLKOWICZ

One must stress a point that certain young historians have found difficult to grasp. Virtually all the basic ideas and philosophies about nuclear weapons and their use have been generated by civilians working quite independently of the military.

BERNARD BRODIE

Admiral Lewis Strauss, one-time head of the Atomic Energy Commission, relates how he witnessed "a meeting at the Walter Reed Hospital, where, gathered around a patient's bedside, and attentive to his last words of advice and wisdom were the Secretary of Defense, his deputies, the Secretaries of the Army, Navy, and Air Force, and all the military chiefs." The central figure in this striking tableau was "a young mathematician who but a few years before had come to the United States as an immigrant from Hungary." The patient was John von Neumann, a defense intellectual par excellence, and Admiral Strauss was moved to confess that he "never witnessed a more dramatic scene or a more moving tribute to a great intelligence."[1] This scene is particularly noteworthy in light of the century-long American tradition of anti-intellectualism and deep disdain by the Establishment for the nonpragmatic and "nonserious" intellectual outsiders. This chapter examines the remarkable rise in the halls of power of a group of intellectuals and experts who were involved in the development of the theories and policies of nuclear war in the United States after World War II; it attempts to account for the sources of their great influence and to consider the broader implications of their rise to power.

INTRODUCTION

One of the most profound, and potentially dangerous, disparities in Soviet and American approaches to military and nuclear strategy lies in their fundamentally dissimilar historical approaches to the relationship between politics and war, that is, the uses of military power for given political objectives. As far as military strategy goes, "most Americans are neo-Jominians and neo-Uptonians,"[2] whereas the Soviets remain followers of the teachings of Clausewitz and of the neo-Clausewitzians among the Marxists and Leninists.

Baron Antoine Henri de Jomini, one of the most influential military strategists of the nineteenth century, maintained that the study and conduct of war was a science that could be reduced to fixed rules and mathematical formulas. He believed that although technologies and techniques of war might change, its "principles are unchanging." This notion, that the study and conduct of war are a kind of science, derives also from the spirit of the Enlightenment and the positive, optimistic American (and English) tradition of democratic liberalism, which envisaged man's ability to control, manage, and order society and forces of conflict by rational, scientific, and technological means. Aspects of this tradition in the nuclear era are embodied in the optimistic premises of crisis management and arms control and in the vast literature on nuclear deterrence strategies. This American fascination with exact control and manipulation of levels of violence, with computer models and complex scenarios of conflict, with the quantification of defense and war management, with war as a rather bloodless game involving largely money, machines, and management techniques, resulted in the development of nuclear war strategies that embodied, according to Walter Lippmann:

> The perfect fulfilment of all wishful thinking on military matters; here is war that requires no national effort, no draft, no discipline, but only money and engineering know-how, of which we have plenty. Here is the panacea which enables us to be the greatest military power on earth—without investing time, energy, sweat, blood and tears.[3]

Although these early post–World-War II views of American attitudes to nuclear war have been somewhat modified by the eventual loss first of atomic monopoly and then superiority, followed by the evolving nuclear stalemate, the underlying basic beliefs have not changed very much: The requirement for the effective punishment of an aggressor, the unambiguous and certain disabling of an offender with minimal ease, and at minimal cost, is considered as valid today as it was in the immediate post–World-War II period.

The other major distinguishing characteristic of American strategic thought derives from the separation of strategy not only from politics, but also from the serious consideration of adversary interests, values, and motives. This follows the tradition of General Emory Upton, who believed that "war and politics are diametrically and fundamentally different." This thesis, developed more than a century ago, had a profound effect on American military thinking, as expressed in an Army War College text that categorically asserted that "politics and strategy are radically and fundamentally things apart. Strategy begins where politics end [sic]." Americans grew up to believe that war is not a concern of political leaders and citizenry, but of military experts alone. As Michael Howard observed, "nuclear weapons have deepened this belief: too complex for the average citizen to comprehend, too awesome even to contemplate, war in the nuclear age has been perceived by many to be something done by 'experts.' "[4]

Soviet strategic thinking has different historical roots; central to their strategic theorizing and military planning is the influence of Carl von Clausewitz and the neo-Clausewitzians in the Marxist tradition, including Marx, Engels, Lenin, Trotsky, and Stalin. War and strategy, in their Clausewitzian conceptualization, did not constitute a science or a body of timeless principles. Clausewitz "rejected both the optimism and dogmatism of the eighteenth century theory" and held that war was neither a scientific game nor an international sport, but an extreme act of violence: "We do not like to hear of generals who are victorious without the shedding of blood,"[5] because the resulting attitude leads to an underestimation of the terrible nature of war and could result in a false sense of security and control over events. Clausewitz was widely read and admired by the Marxists and by the Bolsheviks. Friedrich Engels, probably the most sophisticated student of military affairs among them and known as the "Red General," defined the essential Marxist and Soviet approach to war and strategy, which he derived from Clausewitz and which he bequeathed to the Soviets:

> Fighting is to war what cash payment is to trade, for however rarely it may be necessary for it actually to occur, everything is directed towards it, and *eventually it must take place all the same, and must be decisive.*[6]

This perception of the inevitability and horror of war is also rooted in the Russian historical tradition, as expressed by Tolstoy, who observed that:

> War is the most terrible thing in life, and we ought to understand that and not play at war. We ought to accept this terrible necessity sternly and seriously . . . let war be war, and not a game.[7]

The current position of Soviet strategists on the unsuitability of "eternal principles" of war, and on the need to adapt strategy and doctrine to the given "objective" national and political conditions, goes back to their revolutionary origins and the creation of the Red Army. Trotsky, the founder of the Red Army, cited Clausewitz in arguing that "in practical military arts one should not drive the flowers and foliage of theory too high, one should rather keep them close to the soil of experience." Moreover, in rejecting the notion of strategy as science, he pointed out that despite various claims, "none of the 'national' doctrines of war offered, or could offer, any 'final truth' about war. Each school of thought merely reflected temporary conditions of national existence." He sarcastically criticized those among the Bolsheviks who were carried away by utopian visions of universal, scientific strategic systems. "If we check the 'eternal truths' of military science, we obtain not much more than a few logical axioms and Euclidian postulates. Such principles . . . may well be applied to matters very remote from the art of warfare."[8]

Although contemporary Soviet strategic thought essentially retains this Clausewitzian tradition, it is not ignorant of Western strategic methodologies and concepts. Unlike Western strategy, which largely assumes the adversary thinks as we do, several Soviet research institutes have in the past two decades focused their attention on the studies pursued by American think tanks and military research institutes. On the occasion of the establishment of the Soviet Institute for the Study of the United States of America, several commentators in the Soviet press stressed the "great importance for the comprehensive and concrete study of the economies and politics of the United States of America" as one of the central tasks for Soviet researchers in strategic and international affairs "in whatever field they may be working."[9]

But whereas both American and Soviet research institutes are concerned with strategic and foreign policy studies, the degree of their influence on defense and foreign policy *making* and on the development of strategic theory and doctrine is markedly dissimilar. The influence of United States think tanks and intellectuals was overwhelming and decisive in the immediate post–World-War II era and has continued to a lesser degree over recent years; the impact of Soviet research institutes remains marginal and is largely confined to public relations—serving as host institutions for visiting foreign VIPs— and to research and publication of technical and "scientific" studies of capitalist systems. The Party apparat and the military fully dominate the Soviet defense and strategic policy process and jealously guard their prerogatives, keeping the defense intellectuals and their think tanks at arm's length.

It is important to inquire into the origins and implications of these striking dissimilarities between the American and Soviet approaches to strategy and war and their relationship to politics. Over the years, the Soviet Union and the United States have accumulated virtually endless quantities of data about each other's military technologies, economic and technical capabilities, and the quantities and qualities of their respective weapons systems. And yet, in the final analysis, we seem to understand very little in regard to what purposes and by what kinds of concepts and values these capabilities are to be used. This chapter suggests that the key explanation for this troubling state of affairs lies in the different approaches to strategy and politics undertaken and rationalized by different kinds of experts in the two countries. In the Soviet Union, the dominant expert elites traditionally entrusted with the development and rationalization of strategic theories and policies have been, and remain, military professionals under the political guidance of the Party leadership. In the United States, on the other hand, in the postwar period, this function became dominated by civilian defense intellectuals, to the virtual exclusion of military professionals, under the ambiguous guidance of an ever-changing political leadership.

Given the basic premises and requirements of mutuality, interdependence, and equilibria in the logic of deterrence stability and mutual assured destruction, such fundamental asymmetries in United States and Soviet strategic theories and doctrines are profoundly disturbing.

INTELLECTUALS AND WAR: HISTORICAL PERSPECTIVES

From time immemorial, intellectuals have stood by the side of political and military rulers and offered their advice and legitimating ideologies for the ruler's authority. Even in the earliest civilizations of Mesopotamia and China, the scribe-intellectual played an influential role in contact with the people wielding political power, both in the palace and in the bureaucratic *Great Organizations*. In Mesopotamia, the ruler "in order to establish his rule firmly needed not only the military power . . . but also the bureaucratic know-how . . . and equally important, the royal ideology that was nurtured and elaborated on by scholars and intellectuals who catered to his needs." The special expertise and services offered by these experts-scribes gave them important positions at the court "which offered them not only economic security, but also the reward of special status," which, in turn, "fostered the development of the kind of abstract scholarship reaching into the purely abstract realm of mathematical astronomy."[10] The scribe-intellectual's activities became essential for the proper functioning of the bureaucracies and of the military.

In *The Twilight of Authority*, Robert Nisbet states that from "at least the time of the ancient Greek Sophists, there has been a striking affinity in the West between war and its symbols, roles, and objectives, and a substantial part of the intellectual class."[11] He further asserts that "an interest in war and in the uses of war and of the military is a vivid characteristic of the intellectual mind." Nisbet suggests that there are several reasons for this. Crisis and war, he argues, not only "bring a relief from the tedium and fixity of routine and the monotony that so often characterize life for all but a minority," but also are associated with important and desirable objectives: progress in the sciences, technology, and arts; community of men under stress and danger ("one of the great appeals of war in the modern world has been its capacity to effect moral crusades on the grand scale"); and revolution, which "is allied psychologically to war."[12] He asserts that the attraction of intellectuals to crisis can be seen recurrently through Western history: "The desire to be close to the seat of power, to seek with every means to expand that power, to extend it over as wide a field, domestic and foreign alike, as possible, was as strong in the Rome of Augustus as in the Washington of Kennedy." Above all, he says, there is "the intoxication in participation in great decisions."[13] This view of the intellectual is shared by other people, including Raymond Aron, who maintains that many intellectuals are drawn to war and revolution not because it promises to reconcile mankind or solve the riddle of history, but because violence itself attracts and fascinates more than it repels.[14]

The intellectuals' attitudes toward war are also linked to their affinity and support for the secular, centralized power of the state. Arguing that the state originated in circumstances of war, and that the trappings and instruments of the state evolved essentially in order efficiently to maintain and support large military establishments and war policies, analysts go on to assert that "the state in the West has attracted to itself what can only be called a clerisy of power . . . a long succession of philosophers and intellectuals from Greek and Roman times to our own who have made the political state the temple, so to speak, of their devotion."[15]

Edward Shils suggests that American intellectuals tend to believe that "the state should do it." And although the dominant and most visible tradition of American intellectuals of almost all kinds has been that of distrust toward the regnant political and economic powers, it has, at least since the 1880s, "been associated with a tradition which asserts that the ills of society can be cured by a strong and virtuous authority." He therefore concludes that at the root of this new tradition "was a strong state" and that "a strong state was their ideal."[16] Max Weber saw the bureaucratization of society and the state as dictated largely by the needs "of a standing army and the development of finance associated with it" and that, therefore, the needs of "the disciplined modern army brought about the triumph of the modern state."[17]

An interest in war and in the military is characteristic of the intellectual mind in Western history: "Politically-minded intellectuals have been perforce military intellectuals in substantial degree" Nisbet suggests, and he enumerates a long list of such Western intellectuals ranging from Plato[18] to our own contemporaries. De Tocqueville observed that "men of centralizing genius are fond of war, and men of military genius are fond of centralization in government."[19]

THE GOLDEN AGE OF NUCLEAR DETERRENCE

It is suggested, therefore, that intellectuals in Western societies have often been attracted to the military and war for diverse reasons, and that at several important junctions of history they have made fundamental contributions to the study and conduct of war—in defense of their countries or in the pursuit of an idea. Probably the most significant involvement of intellectuals in the affairs of state and politics has come about in this century, through the transformation of clerks into laymen and intellectuals into experts. This came about through the demands of the growing state and governmental authorities, confronted with the manifest complexity of management of government and society, and the expanding interdependence of political, economic, technological, and military factors in the management of social planning and national government. This rising demand for expertise, and thus for the involvement of intellectuals in the affairs of state and war, came about primarily as a result of national crises, both economic and military. In the United States, World War I and the Depression saw an influx of intellectuals and experts into positions of influence and authority. However, although the rapprochement between the intellectuals and the people barely survived World War I and ended in a bitter split, and although the acceptance of intellectuals by New Deal leaders remained constrained, the intellectuals' ascent into influence and power after the crisis of World War II was pervasive, striking, and seemingly irreversible.

The emergence of this historically marginal and at times alienated and patronized social class into positions of great influence and power is indeed remarkable, particularly since the traditional and popular American view of the intellectual is not flattering. The long national tradition of anti-intellectualism in America was reflected over and over again in the contemptuous and even hostile attitudes of politicians, journalists, and populists of every kind who jeered at them as "men untainted with any practical experience . . . amateurs, dreamers, theorists,"[20] or worse. Adolph Berle, one of the intellectuals involved in shaping the New Deal, recalled that "a politician who talked to a professor in those days kept it a dark secret. We were freaks, like phrenologists."[21]

However, times and styles changed in the decades after World War II. Intellectuals were not simply invited into the corridors of power, they were hailed and embraced by their traditional antagonists and detractors. Journals such as *Fortune, Business Week*, and *Life*, the voices of the business community and practical men of affairs, now perceived the intellectual in a radically different light:

> Today's great national security problems involve an interaction of scientific, technical, economic, and political factors. They demand a type of abstract thinking that is both disciplined and inspired, a type of thinking that has always been associated with a very unmilitary type of person. For lack of a better name, this sort of person usually is called an "intellectual."
>
> For many it has been a shock just how important the intellectuals have become—especially in a country that has always maintained at least a pose of anti-intellectualism since frontier days. The acceptance of responsibility has brought a no-less dramatic change in the intellectuals themselves. Traditionally they have been outsiders, social critics, idealists. But now, they are insiders and forced to come to grips with reality.[22]

Having proclaimed the intellectuals as insiders and serious men of affairs, the establishment went out of its way to court them and to endow them with the most admirable qualities: "The intellectuals who are now playing so important a role in national security seem a new breed. This breed has already made an important contribution to protecting the nation's security." At times, it seemed the very survival of our nation depended on these new Prometheans: "If we get through these nightmarish years with our free institutions intact—and our people alive—much of the credit will belong to them."[23]

The radical changes in the roles and perceptions of the intellectuals in American politics and national security affairs were described by Theodore H. White, the historian-laureate of the American establishment. He wrote in *Life*, in 1967,[24] that there is a "new power-system in American life, a new priesthood unique to this country and this time, of American action-intellectuals. In the past decade, this brotherhood of scholars has become the most provocative and propelling influence on all American government and politics, and their ideas are shaping our defense and guiding our foreign policy." As the world outruns its comprehension of itself, "inherited knowledge and tradition no longer grip unto reality." Yet governments must have solutions to pressing policy problems, and therefore "with almost primitive faith the American government has turned to the priesthood of action-intellectuals," to those men who believe "they can chart the future of our country." For these intellectuals, "now is the Golden Age, and America is the place." Courted by politicians, the media, and business leaders, these intellectuals have in recent years "reshaped the basing and strike patterns that deploy SAC bombers around the globe; led to the great nuclear test ban treaty of 1963; reshaped our economy and . . . made us rich. In the next 10 years, their theories will change America even more."

White has captured the tone and the Zeitgeist of this Golden Age of the action-intellectuals and academic strategists:

> The Academic Strategists have defined the problems of nuclear strategy and developed its vocabulary. They have fostered two revolutions in strategic thought (the limited war revolution and the vulnerability revolution) and they stand on the verge of a third.

This remarkable rise to power of the academic intellectuals was paralleled by a corresponding decline of the group that had traditionally framed American military thought—the professional officers. The revolutionary impact of nuclear technologies and the radical implications for the theories and strategies of war "put a premium on the special forte of the intellectuals theoretical calculations." It also challenged the relevance of military experience—the stock-in-trade of the generals and admirals—and led to a "bitter battle between pen and sword."

The defense intellectuals were at times endowed by their admirers from the business, military, and political communities with Supermen qualities, hailed as men who gave America, at the high tide of her history and world influence, the "capacity for employing our power with wisdom, firmness and restraint which stands at an all-time high." And lest some critics might have wished to taint these men with suspicions of being excessively "interested in military affairs" or evoking some sort of "academic bias against their 'unrespectable' pursuit of research and policy recommendations concerned with war and politics," their eager admirers pointed with pride to the very Pantheon of the defense intellectual community and asserted that "Rand appears to be one of the best investments the United

States government has made since the Louisiana Purchase"[25] and "if Rand did not exist today there would be a most compelling reason for creating it."[26]

The new intellectual elites became the articulators of theories and policies of nuclear war, of international policies; they became the creators of theories of deterrence, limited war, and arms control; they became managers of the defense establishment, of vast budgetary resources, and of scientific–military establishments. They became insiders, even celebrities. How can one account for this new role and influence of the intellectual?

In studying the French Revolution,[27] de Tocqueville was struck by the important role the intellectuals played in it. How, he asked, did "men of letters, without wealth, social eminence, responsibilities or official status" almost alone "speak with accents of authority?" His answer was threefold:

- First, they were almost without political experience and were consequently made the bolder in political prescription;
- Second, their very ignorance of, or disdain for, experience and history made them welcome . . . simple elementary axioms concocted by individual reason broadcast in seductive language made the intellectual class popular with certain men of money and power;
- Third, the corruption and considerable displacement of the old aristocracy in France left a vacuum.

It would not be improper to characterize the rise of the intellectuals in the United States after World War II into positions of authority and influence in the foreign and defense policy establishments as similar to that described by de Tocqueville at the time of the French Revolution. The defense intellectuals were largely devoid of political experience: most of them came from university faculties, research institutes, and large bureaucracies, or from abroad. Their political and military inexperience freed them to announce simple "elementary axioms concocted by individual reason" and then to broadcast them in the "seductive language" of delicate balances of terror, mutual assured destruction, counterforce–countervalue targeting doctrines, limited war, escalation dominance and control, deterrence by denial or by punishment, and so forth—to develop a veritable science of war. It would not be incorrect, in following de Tocqueville's analogy, to say that the very success of the defense intellectuals was made possible by the displacement of the traditional strategic elites in the United States, namely the military professionals who were overwhelmed by the exotic new weaponry, and the strange logic of the new strategic theories, and by the unorthodox demands made on them by political leaders who early on perceived that nuclear weapons were not very usable in traditional ways, and that their primary, or even sole, purpose would not be in their military application but in their deterrence potential. The military, trained for generations to plan for combat, attack, and victory, were not prepared for this new condition, which proscribed victory as impossible and undesirable, and where permanent stalemate and compromise with the deterred enemy were the central objectives. The defense intellectuals, however, clearly understood; they wrote the rules of the new game of victory by stalemate and by confrontation avoidance. This radically new purpose for nuclear weapons was defined in a celebrated axiom of Bernard Brodie who, at the very dawn of the nuclear era, asserted: "Thus far, the chief purpose of our military establishment has been to win wars. From now on its chief purpose must be to avert them. It can have no other useful purpose."[28]

In the post–World-War II period, after an initial period of nuclear monopoly and strategic superiority, American leaders were suddenly faced with an unprecedented vulnerability to potentially devastating enemy attacks as the credibility of the massive American retaliatory umbrella became questioned. Moreover, the threat of nuclear proliferation and the renewed political activism of the Third World introduced an increasing complexity to international politics and decision making. Faced with these new developments in technology, strategy, and politics, American leaders searched for ideas and policies that would accommodate these changes, stabilize the balance of terror, and ensure the security and survivability of the nation. In this pursuit, they eagerly sought the advice of the newly emerging groups of intellectuals-experts in the strategic and technical disciplines.

Several groups of intellectuals-experts welcomed into the foreign and defense policy bureaucracies of government have since had an overwhelming impact on Western theories and doctrines of war, on the uses of military force for political purposes, and on the techniques of efficiently joining economic means to given policy ends. These groups shared a professional and intellectual involvement in the problems of war and politics in the nuclear context. They generally believed that these problems lent themselves to systematic study, that international politics contained a certain causal logic, and that predictable results could be obtained in the international arena by the application of different kinds and levels of threat and coercion. These new elites[29] came largely from academic and scientific backgrounds and were often associated with large universities, research institutes, or government bureaucracies. They rejected traditional solutions for complex political–military problems and instead preferred an approach of maximizing policy objectives within a relatively value-free context. The civilian defense intellectuals—most of whom came from the disciplines of economics, mathematics, and the social sciences—based their analyses and policy prescriptions concerning strategic theory and decision making on an assumption that since the strategic and technological context of the nuclear era was radically changed, it was necessary to "forget the past, return to first principles and attempt to build up a theory appropriate to the new age." They subsequently developed a pure theory of strategy with premises, logic, and conclusions relevant to East–West relations, but not derived from them. Their influence derived from their ability to provide a rationale for a policy of stable deterrence based on secure second-strike forces, and it led to policy recommendations determined largely by technology. They based a large part of their reasoning on game theory, which provided "a means of reducing strategic problems to a manageable form in which the dilemmas and paradoxes of the age could be bared and solutions explored."[30] As a form of logical reasoning, it depended on premises that were subjectively determined and that could offer valuable insights into behavior. What it could not do was predict how actors in the real world would respond to situations approximating those in the theoretical models, or even say how they ought to respond.

The defense intellectuals radically redefined the traditional ideas and policies of war and strategy in the United States:

- Nuclear strategy as a science. The logic, dynamics, and management of nuclear war and its deterrence can be explained and controlled by precise, quantifiable methods and policies involving the threat of the use of force in order to affect the adversary's decision.
- Nuclear war as an abstraction. The study and decision making regarding nuclear war and deterrence strategies can best be accomplished by means of abstract concepts from game theory, economics, and the reification of the "real world" by crisis scenarios clustered around the notions of deterrence stability, escalation control, damage limitation, and calculations of assured destruction.
- Strategies of nuclear war and deterrence as apolitical and ahistorical concepts. Only "hard," quantifiable variables of military capabilities, technological parameters, and economic resources are relevant in the calculus of nuclear war and deterrence strategy. Such "soft" variables as historical patterns, cultural traditions, political practices, and personality traits are not particularly relevant because they cannot be accounted for in meaningful analytic terms.
- Nuclear war and strategy as a management problem. The central concern for the military and strategic decision makers entrusted with

national security is almost exclusively to be focused on the efficient application of available resources to the given ends provided by the political leadership. Thus, war and strategy become primarily a problem of the most efficient management of finite resources, the observance of cost-effectiveness and control standards, and the adherence to sound planning and managerial rationality.

The central theoretical–methodological concept for the management of this nuclear strategic system derived from the deterrence paradigm, which was developed by the defense intellectuals in the 1950s and 1960s. At the root of this key concept were several fundamental assumptions:

- Nuclear war is not "winnable" because the likely cost (death and destruction) would exceed the likely gains. Thus, rational calculations by rational decision makers will inevitably lead to war avoidance, particularly if both sides are aware of the irrationality of launching nuclear strikes against an equally strong nuclear adversary.
- War avoidance necessarily becomes the overriding objective and purpose of the deterrence system and is to be achieved by means of credible deterrence posture and survivable retaliatory forces. That, in turn, presupposes the need for the maintenance of sufficient capabilities and of high deterrence credibility, and for a firm and clearly communicated resolve to use these capabilities in the event of need.
- The model of deterrence stability, the key objective of deterrence policies, and its management derive primarily from game theory, which prescribes the proper strategies for rational decision making in conflict/cooperation contexts where the acceptable, though not necessarily preferred, outcome depends on tacit bargaining and cooperation by the players, and where winning often involves the acceptance of the best-of-worse outcomes, that is, mini-maxing strategies.

Stripped of its cumbersome language, the deterrence paradigm appears to be grounded in an updated version of the "normal science" of the economics of Adam Smith, modified by the notions of hedonistic utilitarianism of Jeremy Bentham and his "felicific calculus" and strategies of mini-maxing, and further modified and adapted to the stark realities of nuclear weaponry and the nuclear balance of terror. The deterrence stability paradigm is predicated on the need for balance and equilibrium in the strategic relations between the hostile systems. It is further predicated on the belief in the ability to control, manipulate, and stabilize crises and conflicts by the discrete application of sanctions and gratifications, by threats to inflict pain or promises to bring pleasure—a veritable modern "unfelicific calculus" wedded to Adam Smith's notion of a dynamic equilibrium of market forces, where selfish drives and interests collide and become transformed and stabilized into common and public good through the "great and immense machine, whose regular and harmonious movements produce a thousand agreeable effects" and where, therefore, "the self-interest drives civilized society in harmony."[31]

Like the earlier inventors of the ideal economic man, the contemporary inventors of the perfectly rational strategic man are vulnerable to criticism regarding the connection between the abstract models and sociopolitical reality. Thorsten Veblen's critical observations concerning the classical economists and utilitarians are remarkably relevant in the contemporary context:

> This perfect competitive system, with its untainted "economic man" is a feat of scientific imagination, and is not intended as a competent expression of fact. It is an expedient of abstract reasoning . . . but as happens in such cases, having once been accepted and assimilated as real it becomes an effective constituent in the inquirer's habit of thought and goes to shape his knowledge of facts. It comes to serve as a norm of substantiality or legitimacy.[32]

This tradition of the classical economists and utilitarians, which links directly with the inventors of the modern deterrence theories, based on concepts of game theory and of stability and harmony among competitors and adversaries, shares a perception of the rational economic–strategic man: the decision maker as a "lightning calculator of pleasures and pain" who has "neither antecedent nor consequent," one who "oscillates like a homogenous globule of desire and of happiness under the impulse of stimuli," and who is an "isolated, definitive human datum in stable equilibrium."[33] In short, the classical economists and utilitarians and their modern deterrence theorists premised their theoretical systems on a "doctrine of the harmony of interests" in a beneficient competitive and antagonistic system.

This intellectual legacy is clearly evident in the works of the inventors of the modern nuclear deterrence system, and its implications for political and military realities are profoundly troubling. Western defense intellectuals have invented virtually all of the conceptual foundations of nuclear deterrence theory. They have defined the theories (of massive retaliation, limited war, escalation control, assured destruction, flexible response, escalation dominance, etc.), the methodologies (from game theory, systems analysis, rational actor models, etc.), and the language and rules of the game for rational strategic decision making for superpower interaction in the nuclear context, which is focused on the objective of deterrence stability. They have also largely dismissed Soviet theoretical, analytical, and policy-prescriptive approaches to these issues, finding them methodologically unsophisticated, theoretically primitive, lacking in quantitative precision, and laden with ideological irrelevancies. Thus, American and Soviet approaches to strategy and war remain in many ways profoundly out of sync.

SOVIET–AMERICAN STRATEGIC DISSIMILARITIES

The basic differences that separate Soviet strategy from that of the United States "mainly reflect dissimilarities between the Soviet and U.S. strategic cultures, not any inherent incapacity on one side or the other to appreciate military logic." While United States strategic concepts are "largely the artifacts of civilian decisionmakers who have set the drumbeat of U.S. national security over the years . . . by contrast, in the Soviet case those responsible for formulating doctrine and strategy, setting R&D requirements, selecting weapons for deployment, and sizing the forces all wear uniforms."[34] In general, the Soviets have shown no attraction to the "intellectual baggage" that has been associated with United States defense planning throughout the postwar years. It is, therefore, in their approaches and understanding of strategy, of the relations between war and politics, and in the uses of force for given ends that the two countries so strongly differ.

Although the objectives of Soviet and American strategic nuclear policy are superficially similar, each country has pursued different policies based on differing assumptions regarding the very nature of political rationality and military utility of forces, deployment postures, and the controllability of nuclear war. Although both countries have embraced the concepts of deterrence of nuclear war and the limitation of damage if deterrence should fail, and although their military technologies and operational concepts are rather similar, there is a wide gap in their fundamental military doctrines and strategic theories and in their perceptions of the relationship between politics and war. Since the early 1960s, the overriding objective of United States strategic nuclear policy has been the development of a strategic posture designed to enable the United States to control any nuclear exchange in order to limit damage at the lowest possible levels while ensuring that the outcomes are favorable to the United States. The notion of controlled response, which was developed by the Kennedy administration in 1961–62 and which governed the design of SIOP

63 (the single integrated operational plan, or the plan for general nuclear war), has been largely retained and refined by successive administrations. The central theme of these strategic policies has been that the limitation of damage in nuclear war can best be achieved by controlling escalation so as to terminate any nuclear exchange at the lowest possible level and, insofar as possible, to give the Soviet Union every incentive to avoid escalation to major urban–industrial attack targets.

There is no exact parallel to any of this in Soviet strategic thinking and nuclear war planning. On the contrary, Soviet strategic planners believe that the best approach to limiting damage to the Soviet Union is the rapid and wholesale destruction of the ability of the United States and its allies to wage nuclear war. The Soviet view of deterrence, unlike that of the United States, involves "neither the notion of assured destruction or unacceptable damage, nor that of limited or controlled nuclear options. Rather, deterrence of nuclear attacks is best achieved by the ability to wage a nuclear war successfully."[35] In other words, the better the Soviet forces are equipped and prepared to fight a nuclear war, the more effective they will be as a deterrent to a nuclear attack on the Soviet Union. And, in the event that deterrence should fail, these forces will then be used purposefully and massively for military victory.

The Soviets vigorously reject the key Western strategic concept of mutual assured destruction as a consummate abandonment of leadership responsibility. Soviet strategists and defense planners have consistently and repeatedly rejected such Western approaches and theories of nuclear war and their strategic prescriptions and policies. They find Western strategic sophistries objectionable and unacceptable because they assert that, in the West, key political decisions are often subordinated to narrow technological imperatives on the one hand, and to the formal logic and abstract notions of game theory and the formulae derived from the concept of mutual assured destruction on the other. A recent interpretation of this Soviet position argued that:

> American strategic thinking—born predominantly of civilian defense specialists bearing legal, technical and distinctly non-military intellectual outlooks—is deeply rooted in the proposition that nuclear war is unwinnable in any practical sense.[36]

This intellectual and civilian influence on American strategic policy making has produced an increasingly predominant belief that deterrence stability (hence U.S. security) is best served by a strategic environment of mutual vulnerability. The Soviets reject mutual vulnerability out of hand as an abdication of political responsibility.

Western civilian strategists show their exasperation with what they describe as unsophisticated, primitive, and even obsolete Soviet strategic thinking, pointing at the most influential postwar Soviet work on nuclear strategy, edited by Marshal V. D. Sokolovsky,[37] and complaining that "nowhere in this book, as in most Soviet literature as well, are there to be found signs of serious professional interests in concepts like controlled response and restrained nuclear targeting" that have been widely accepted in the West. Moreover, they continue, such unsophisticated and primitive Soviet strategic thinking tends to "discourage the United States and its allies from entertaining ideas that ground rules of some sort might be adopted for limiting the destructiveness of a war, should one occur."[38]

At times, Soviet and American approaches to strategies of nuclear war and its deterrence seem to be a dialogue of the deaf. The American strategists assert that their theories, methods, and policies are the only correct way for the maintenance of nuclear deterrence, and they have trouble understanding why the Soviets do not agree with them. However, the Soviet strategists and military leaders see the problem in a radically different light: "The idea of introducing rules and games and artificial restrictions by agreements seems illusory and untenable. It is difficult to visualize that a nuclear war, unleashed, could be kept within the framework of rules and would not develop into an all-out war. In fact, such proposals are a demagogic trick."[39]

Soviet strategists ask how "is it possible to set the rules of the game, say to agree not to bomb peaceful towns, to limit the power of the bombs to 10 kilotons, and use them in battlefield only, to ban all explosions except air bursts with no radioactive aftermath, etc."[40] The noted Soviet strategist General N. Talenskii succinctly expressed the Soviet unease with the Western formulations of nuclear strategy: "When the security of the state is based on mutual deterrence with the aid of powerful nuclear missiles, it is directly dependent on the good will and designs of the other side, and that is a highly subjective and unreliable factor."[41] And Soviet Defense Minister Grechko, while pointing out that Western military theorists advocate views that differ profoundly from those of Communist strategic theorists, asserted that the basic difference lies in the fact that Western strategists "consider war as a mere armed clash between the two sides" and therefore "emasculate the political content of the concept of war."[42]

The Soviets regard such entrenched Western concepts as demonstration attacks, limited nuclear options, slow-motion counterforce duels, and so forth, with alternating bemusement, perplexity, and sarcasm. They find such ideas questionable and lacking in credibility because, as they assert, they would have to "rely on the chance that the aggressors will be prudent and will impose certain limits on the use of nuclear weapons."[43] In any case, the Soviet analysts go to great lengths to explain that American and Soviet strategic theories and policies are not comparable because their histories, defense requirements, and political goals are dissimilar: "The Soviet Union strengthens its defense capability to suit its own specific development of science and technology, its geographic situations, the experience of past historical periods in building up its armed forces."[44]

DECLINE OF DETERRENCE: THE REAGAN ADMINISTRATION AND MAD

The legacy of the defense intellectuals persists in the United States defense and foreign policy establishments; however, the centerpiece of their strategic theoretical edifice, the concept of mutual assured destruction (MAD), has been attacked from all directions:

- Critics from the right argue that MAD does not work well because it failed to constrain the Soviets, who persist in their aggressive designs. They maintain, however, that deterrence can be made to work by means of technological fixes like the Strategic Defense Initiative and general modernization and augmentation of our defense capabilities.[45]
- Critics from the left maintain that deterrence, and particularly MAD, does not work at all. It simply provides a deterrence ideology that demands ever higher levels of costly arms racing in the name of the vaunted goal of deterrence stability, without any visible increases in national security or international stability. These critics see possible solutions to this dilemma in political and unilateral arms control and disarmament actions.[46]
- Even supporters of deterrence and MAD, who see no viable alternatives to them, are concerned with the inevitable security dilemma and ill-advised technological fixes like Star Wars (SDI). They therefore counsel more prudence and restraint and greater commonality of interests among the adversaries, to be implemented by arms control agreements and by the kinds of arms acquisition and deployment policies that would reduce perceptions of threat and enhance stability.[47]
- Specialists in foreign and defense policies who consider the domestic and bureaucratic politics in each country to be the primary deter-

> minants of defense and foreign policy decisions, rather than the inputs and threats from outside these political systems, see President Reagan's decision to proceed with SDI as being largely motivated by partisan and domestic concerns, rather than by escalating Soviet threats or determinants of strategic theories.[48]

Still others perceive the main motive force behind United States and Soviet strategic theories and policies as stemming less from the real world of superpower confrontations, than from the differing historical, political, and philosophical traditions in each country, and as having been powerfully influenced in the Soviet case by their unique, tragic, war-torn history and the sociocultural traditions of fatalism and *kto-kovo* distrust of claims to political rationality and benign motives of adversaries in conflictive situations.

And, of course, the most vocal and powerful criticism of MAD deterrence theories and policies of recent decades has emanated from the Reagan administration and its supporters. The Reagan administration has made extravagant claims to having ended the years of political neglect and theoretical strategic morass of its predecessors, claiming to have brought about a revolution in the defense policies and strategic thought of the United States: "The world has changed profoundly since the 1950s and early 1960s when most of our conceptual arsenal was formulated" asserts Defense Secretary Weinberger, who finds that most of these "concepts are now obsolete" since they were associated with an "era that has vanished." He defines these concepts of mutual assured destruction as "a mutual suicide pact" and instead, he asserts, the Reagan administration offers the alternative of a moral and effective deterrence in which the threat of escalation could be controlled, collateral damage minimized, and the lethality of war reduced, and in which the "impossible dream" of nuclear missiles becoming obsolete would become a reality.[49]

On closer examination, however, these extravagant claims to a strategic revolution seem less impressive than its authors assert. Their main claim lies in the rejection of the immoral and unworkable strategies of consensual vulnerability[50] and the mutual suicide of MAD, and in their proposals to render nuclear weapons obsolete and impotent through SDI. Moreover, they triumphantly claim that these policy initiatives have already brought about success, having persuaded the Soviets to moderate their aggressive behavior and to return to the arms control negotiating tables, where they have presented seemingly serious proposals for substantial reductions in strategic forces. Several observations are warranted here.

First, regarding the claims to revolutionary changes in United States defense and strategic policies of the Reagan administration, a number of important shifts have taken place in strategic thinking and policy making in the past decade, becoming formalized and emphasized under the current administration. The interesting point about these changes lies less in their alleged innovative or even revolutionary aspects than in the fact that they have been around and quite well known for decades; they have also for decades been a basic staple of Soviet strategic thought and military policy. We appear to have borrowed or adopted them from the Soviets, after decades of patronizingly dismissing their strategic theories and doctrines. It is rather ironic that this kind of wholesale borrowing or adaptation, with United States strategists and political leaders, consciously or otherwise, paying the Soviets the ultimate compliment of imitation, should take place largely during the Reagan administration.

What kinds of strategic ideas did we rediscover or borrow from the Soviets, which have since been incorporated into United States strategic theories and defense policies?

- Strategic defense is good and not destabilizing, and not necessarily threatening to deterrence stability;
- War-fighting capabilities, postures, and doctrines are not destabilizing, but rather deterrence enhancing, contributing to deterrence stability;
- Deterrence by denial, rather than by punishment, makes good sense and serves to enhance damage limitation;
- Decapitation strike concepts and policies are not counterproductive and do not necessarily complicate the problems of intrawar deterrence, control of nuclear war, or war termination. They simply enhance the credibility of deterrence by threatening the highest values in the Soviet Union, namely, Party and military leaders, command and control centers, important government establishments, economic resources, and so forth;
- Force modernization and development across the whole spectrum, including the policy of escalation dominance, makes very good sense, particularly in the assignment of important roles and missions to nonstrategic forces.

These and other defense policy and strategic theoretical innovations claimed on behalf of the Reagan revolution are not really new. They have been a commonplace in the Soviet defense system and have merely become Americanized.

Second, although the authors of the "new" strategic theories and policies vehemently reject any connection or indebtedness to their predecessors, who are said to have sinned against United States traditional morality in war and to have permitted the Russians to get by far the best deal in the strategic balance and arms control negotiations, they appear on closer examination not really to have abandoned the theoretical and strategic legacies of their predecessors. To be sure, MAD is vigorously rejected, and the earlier strategic notions that vulnerability equals stability equals security have been roundly condemned. However, some of the most fundamental premises and assumptions of the MAD era seem to have survived, although unacknowledged: the notion of the rational–unitary actor; the centrality of the concept of deterrence stability, augmented by the corollaries of deterrence credibility and clarity of communications; the concept of the controllability of escalation and of nuclear war; and, in the final analysis, the utility of arms control as an auxiliary to stable deterrence, and so forth. These basic conceptual and theoretical foundations of the MAD crowd remain embedded in the current revolutionary strategies.

Third, claims of cause and effect have been made by the current administration regarding Reagan strategies and policies and their desired impact on Soviet leaders: they suggest that the administration's policies have impressed, sobered, and possibly intimidated Soviet leaders sufficiently so that they have moderated their aggressive and arrogant ways stemming from their militarily superior position, and have now become persuaded of the need to negotiate with the West. How tenable are such claims?

It is quite possible that such claims are true, and, if so, they would provide evidence of the success and wisdom of the Reagan strategic revolution. However, one could possibly devise other explanations and scenarios for these developments, explanations that have less to do with Reagan's *virtu* and more with *fortuna*. Moreover, one could suggest that the changes in Soviet behavior were brought about not so much by deliberate United States policy initiatives but by changes in Soviet domestic politics and in the objective conditions and the correlation of forces in the international arena.

After a long period of conservative, risk-averse, and eventually quite incompetent leadership under the Brezhnev regime, followed in quick succession by the interregnums of Andropov and Chernenko, the Soviet Union got a new, younger, more pragmatic, and non-Stalinist leadership under Gorbachev. This leadership appears to have assessed the sour legacy of the past two to three decades and decided on a shift in national priorities, at least in the short run. The new, pressing priorities appear to include stabilization of the domestic front and a greater attention to the long-denied consumer needs of the restless population; modernization of the economy and of the vast and inefficient governmental–economic bureaucracies; and, stabilization of the brittle alliance system and reaffirmation of Soviet influence and control.

This new, albeit potentially transitory, shift in national priorities was most likely dictated and facilitated by several developments: the Soviet Union has already achieved clear strategic parity and sufficiency, and thus there is less urgency to maintain the very high levels of resource allocations to defense; the modernization of the Soviet economy and technology was postponed and neglected by previous regimes, and it has become urgent to attend to this matter lest the Soviet Union resign itself to becoming a second- or third-rate, underdeveloped, technologically backward system;[51] the Polish *malaise* has shaken up many Communist leaders who have observed that long-suppressed consumer demands can erupt in dangerous, popular uprisings that challenge the Communist Party's ability to cope with them and necessitate virtual military takeover of the Party's governing authority.

It is thus suggested here that the Gorbachev leadership needed a *peredishka*, a breathing spell of time, to attend to pressing demands and needs on the domestic front and in the alliance. Thus, the rationale for a more conciliatory, accommodating, and less confrontational foreign and defense policy posture was presumably based on the assessment of the urgency of these internal developments.

The urgent priorities on the domestic and alliance fronts counseled the need for a more accommodating and friendly policy toward the self-assertive foreign policy of the Reagan administration. It would seem reasonable, under the circumstances, for the Gorbachev regime to strike a pose of reasonableness, to avoid provocation, to appeal to Western public opinion, and to relax Soviet–American relations. Moreover, while giving up little, the Soviets could press the Europeans and Americans on the ever-popular arms control issues, which could rationalize and justify the Party leaders' strong interest in stabilizing their very high defense expenditures and thus serve their urgent internal priorities.

It is not unlikely that the Reagan strategists, like those in previous administrations, saw cause-and-effect relationships in their policies aimed at influencing Soviet decisions and behavior, when the evidence for such a conclusion is marginal at best.

THE DETERRENCE TRAP

The central theoretical concept relevant to the management of nuclear deterrence policies is embodied in the nuclear deterrence paradigm that was developed in the 1950s and 1960s. At the root of this key concept are three assumptions: nuclear war is not winnable; nuclear war avoidance is imperative; and strategic stability is the most desirable condition for the management of war avoidance and war deterrence, and thus for the survival of the deterred countries. This model of strategic stability derives primarily from game theory, which defines the proper strategies for rational decision making under conflict–cooperational conditions, whereby the survival of one depends on the survival of both, and where winning often involves the acceptance of less than optimal outcomes, that is, mini-maxing.

There is a fundamental problem with this deterrence paradigm, which leads in the final analysis to a deterrence trap, created by certain unintended and misperceived consequences of the several interrelated aspects of the nuclear deterrence system as it has evolved over the past few decades. The first problem derives from the perceived logic and dynamics of the nuclear deterrence system; the second problem derives from the methods and policies developed for the management of this system.

The Imminence of War Threat

Strategic nuclear weapons have radically changed the time, scope, and destructive force of war. Since outcomes of nuclear war could be decided during the initial period of war, in minutes or hours, nuclear protagonists in a deterrence condition must be permanently ready to "go to war." They must have all war capabilities on hand since there will be insufficient time to redeploy their forces and to mobilize their resources as in past wars. The nuclear powers must, therefore, operate on the assumption of a permanent imminence of war and, under the logic of deterrence, must be constantly ready credibly to deter and, if necessary, strike the potential aggressor in order to limit their own damage.

Moreover, under the same logic of the deterrence paradigm, the nuclear powers need to maintain a high-readiness, war-fighting posture in order to pursue successfully:

- Credible arms-control policies, following the dictates of bargaining chip, arm-to-disarm practices.
- Properly managed war termination.
- Retention and enhancement of deterrence stability.

These high levels of readiness and war preparedness are mandated for purposes of deterrence, for signaling intent, for credibly demonstrating capabilities, and for communicating threats and sanctions. This situation, however, creates another problem.

The Immanence of War Threat

The very conditions and policies demanded for coping with the potential imminence of nuclear war threat also create a condition of immanence of war threat. The high levels of military readiness, capability, response, and credibility mandated for deterrence and other purposes may create the perception, under certain circumstances, of aggressive intent and thus serve as a permanently immanent war threat.

The Logic and Dynamics of Deterrence Stability

In order to maintain the highly desirable condition of deterrence stability, strategists and defense managers follow the dictates of Soviet–American military and political confrontational relations characterized by intense distrust, threat perceptions, and ideological animosities, emphasized by pressures from domestic and bureaucratic interests. The aggregate of these pressures and demands leads, in the balancing of the deterrence equilibrium, to an overcompensation in weapons acquisition for any perceived advantage gained by the adversary, to an overinsuring against future surprises, and thus to a constant upward ratcheting of forces at ever higher levels of lethality. The factors driving this momentum include the worst case assumption, the action–reaction phenomenon, the mad momentum of technology, the bargaining chips strategy for arms control, and the high and growing levels of research and development. The inevitable outcomes of these factors interacting with each other are a permanent nuclear war system and an inability to return to a predeterrence context. In turn, the permanence of the nuclear war system creates a circular logic that reinforces the dictates of the deterrence paradigm and creates additional pressures for the maintenance of deterrence stability and equilibria at ever higher levels.

The logic of the deterrence system effectively promotes and ensures a constant stimulation of arms development and acquisition and undermines efforts to reduce, stabilize, or control the ever-growing arms race spiral. Thus, under the logic of deterrence:

- Efforts to achieve a balanced deterrence equilibrium stimulate arms racing.
- Efforts at achieving realistic arms control agreements tend to stimulate arms racing.
- Efforts aimed at humanizing nuclear war, at reducing its lethality to civilian populations via counterforce proposals, destabilize the deterrence system and stimulate arms racing.
- Efforts intended to reduce sharply the nuclear capabilities of both sides are considered destabilizing, because, it is argued, there would

be an insufficiency of targets, thus possibly motivating first-strike incentives.[52]

- Efforts aimed at stabilizing nuclear capabilities at high levels are considered destabilizing to the system, because they are likely to encourage the use of force at the lower end of the force spectrum and possibly to motivate some to overcompensate for deficiencies in nonnuclear capabilities by the resort to nuclear force, and so forth.

The logic of the deterrence stability paradigm seems to imply that any significant effort to modify or change it is doomed to defeat: the lowering of lethality of nuclear war via arms control is considered destabilizing; raising the lethality is considered destabilizing; containing the volatility of the deterrence system via arms control is considered destabilizing; the efforts to reduce and limit damage via civil defense, denial strategies, and even Star Wars are destabilizing, as are policies for the increased usability of conventional forces; and the efforts at reduction and better control of escalation momentum (via limited nuclear options) and policies for less restrained control of the escalatory momentum (via massive retaliation) are considered destabilizing. Even attempts at setting up some reasonable rules of the game to govern and stabilize the deterrence relationship (via mutual assured destruction or arms control) are considered destabilizing, the former because of possibly misperceived intent and the latter because of the bargaining chip stratagem.

The deterrence paradigm has become an end game, a trap: we cannot let go of it because we have no realistic ideas how, and with what, to replace it, and *pace* President Reagan, Star Wars remains an impossible dream. In the meantime, both sides go about the "normal science" of managing their respective deterrence strategies, a dangerous and expensive business.

THE LEGACY OF DEFENSE INTELLECTUALS: ASSESSMENTS AND CONCLUSIONS

The assessment of the legacy of the defense intellectuals, the architects of the deterrence system, leads one to rather melancholy conclusions. This "priesthood of action intellectuals," who like Prometheus stole the fire from the sun and sought to control it and to benefit mankind, seems to have earned the wrath of the gods.

There is little doubt that the defense intellectuals improved and rationalized policy making during a most crucial period of recent history. They expanded the universe of strategy and policy and described its complexity while introducing an element of order and predictability into policy making. They also introduced a number of less desirable influences to the policy process:

- Having believed that the adversary was implacably hostile and that he subscribed to the view of politics as a zero-sum game, the elites argued for the primacy of national defense over other societal goals.
- Having accepted the view that policy making in the nuclear context is so esoteric that only specialists can cope with it, they tended to reject public participation.
- Having accepted the idea that war in the nuclear context is "continuation of management by other means," they fell victim to their own illusions by seeing the complexities of international politics as a management problem in which rational means–ends techniques would yield predictable results.
- Having embraced, deliberately or otherwise, a paradigmatic view of politics and war, they mistakenly tended to apply this paradigm to the anarchic complexities of international politics. Overabstraction, scientism, numerology, and technical jargon are all part of the legacy of this professionalism.

Civilian defense intellectuals were highly successful and quite influential during the golden age of the nuclear era. However, their theories and policy recommendations soon encountered skepticism in various quarters. The professional military officers found their excessive claims and intellectual arrogance offensive, and others from the scientific and academic communities expressed concern over their ahistorical and apolitical biases, warning that more was being sought in the scientific method than would ever be provided. Some were troubled by the defense intellectuals' bizarre recommendations based on a pseudoscience that was insensitive to the real attributes of nations and decision makers. This new science of war and deterrence was thought to be "based upon assumptions about human behavior which seem totally unreal. It neither constitutes scientific analysis nor scientific theorizing, but is a non-science of untestable speculations about Western and Soviet bloc behavior."[53]

The golden age of the action intellectuals and academic strategists has passed. The high praise and promise of these Prometheans have dimmed, the intractables of international and domestic politics, of economic constraints, and of unanticipated historical and socioeconomic forces in various parts of the world have jarred the wheels of their elegant designs, their grand theories, exotic scenarios, and authoritative doctrines. Clearly the golden age has passed, a brief period of intellectual ferment and political influence nostalgically located sometime in the decade of the mid-1950s to mid-1960s, when the promise of harnessing and controlling the nuclear Moloch was high. The proponents of the new science of war promised to safeguard national security, to provide a stable international environment, and to develop a rational decision-making system for the management of America's national interests in a hostile nuclear world through a combination of theories of deterrence, arms control, and techniques of crisis management. Analysts with a more jaundiced, or at least a more detached, view of the defense intellectuals' claims have observed that: "In 1961, the promise was high. The civilian strategists came to Washington to assume influential roles" in the development of strategic and foreign policies in the Kennedy–McNamara team. However, by the 1970s, "It is fair to say that their performance had not lived up to their promise. And that's putting it mildly."[54]

In the final analysis, what brought on the eventual decline of the defense intellectuals' influence on military policy was their failure to provide a credible and convincing answer to the question of what to do if deterrence failed: "They offered a measured response as an alternative to all-out war, but the arguments lacked plausibility and collapsed into paradox."[55] And although their excessive use of quantification and their scientific pretensions created an aura of objectivity, the tenuous relationship between the contrived world of their theories and the real world limited the practical validity of any conclusions. Their models could not predict or explain actual behavior, though they suggested criteria for judging behavior. As Hedley Bull observed: "A great deal of argument about military strategy . . . postulates the 'rational action' of a kind of 'strategic man,' a man who on further acquaintance reveals himself to be a university professor of unusual intellectual subtlety."[56]

Others observed that "academic strategists overintellectualize their approach to war and make recommendations far too subtle and complex for any policymaker to apply in the real world"; and, in any event, their whole approach to nuclear war and its deterrence leads to an endless and increasing arms race. Some, like Professor Edward C. Banfield of Harvard, vigorously opposed the idea of intellectuals in positions of power:

> The premium of scholarship for a professor is all too often originality, not correctness. A politician or businessman must pay the price for being wrong; the academic never does. . . . I think it's a national tragedy that people in decisionmaking roles turn over to intellectuals the right to make decisions. And it's bad for scholarship, too.[57]

In his preface to Mannheim's *Ideology and Utopia*, Louis Wirth referred to *kikenshiso*—dangerous thoughts whose knowledge among the populace might subvert the sanctioned beliefs and undermine the existing order. One such sanctioned belief[58] is the notion that the proper and traditional role of the intellectual in Western societies has always been that of the outsider, the critic, the "unveiler," one who is alienated from the authorities and society. Yet, the study of the intellectual groups associated with the United States foreign and defense policy establishment seems to challenge these self-images of the intellectuals and give credence to Paul Hollander's observation that "it is useful to remember that the intelligentsia (or intellectuals) have a much longer history as the upholders of the traditional world views and supporters of authority than as performers of their special role which prescribes the questioning of authority, marginality, and highly developed critical faculties."[59] In any event, the rise of the intellectuals' influence as nuclear strategists, crisis managers, and presidential advisers is not without certain troubling consequences. To be sure, these experts-intellectuals have systematized and routinized concepts and decision making in the complex nuclear system; they have alerted us to certain crucial and dangerous implications in the traditional attitudes toward war and strategy; and they have created an impressive body of theory on war and strategy in the nuclear context. However, they have also contributed significantly to new patterns of theorizing and policy making that are troubling and dangerous and that have resulted in:

- The depoliticization of war and the progressive militarization of politics.
- An excessive abstraction and overintellectualization of warfare and strategy.
- Pervasive beliefs and illusions of control over unleashed forces of war, of the ability to manage scientifically international and nuclear crisis situations.
- Unfounded claims and illusions of the pervasive rationality and cool calculation of decision makers under the extreme stress of nuclear crisis.

Some general conclusions emerge regarding the intellectuals' impact on the theories and policies of war in the nuclear era. Far from having invented a modern science and policy for the management and prevention and deterrence of nuclear war, the American defense intellectuals have in fact, consciously or otherwise, projected an American ideology in the form of strategic theory, an ideology based in part on some of the traditional beliefs and values of our history. Moreover, this ideology, imbedded in the theory of deterrence, simply reflects the inherent preferences and values of the intellectual class and their traditions: an unquestioned belief in rationality, abstraction, logic, quantification, stability, and harmony. This ideology in the form of strategy, despite its claims to universality and scientific objectivity, is intensely ethnocentric and fails to take proper account of the alternative values, traditions, and preferences adhered to by the allegedly interdependent deterrence partner, the adversary, the Soviet Union. This ideology, because of its ethnocentric orientation, violates the very dictates and requirements of the deterrence paradigm—namely, the necessary and unavoidable interdependence among nuclear adversaries in a mutual deterrence relationship, which demands that both parties must cooperate and both must understand the rules of the game of nuclear deterrence, thus binding both equally lest they both perish.

Arthur M. Schlesinger, Jr., observed that revolutions are always made by intellectuals who are sooner or later always replaced by practical men. The American defense intellectuals have indeed revolutionalized the way we think about and plan for war in the nuclear context. Although the influence of the defense intellectuals is waning and is being replaced by that of practical men, their ideas persist in different form. The defense intellectuals have created a theoretical system that "works like a narcotic,"[60] creating a dangerous dependency that is difficult to escape. Moreover, as Fritz Ermarth suggested, in the United States strategic culture, "when it comes to nuclear strategy, we do not believe much in 'real' nuclear war anyway." We are essentially concerned with devising theories and capabilities that are "adequate and persuasive in a peacetime setting."[61]

Theodore White may have provided the proper epitaph on the defense intellectuals in the United States when he concluded his lengthy and enthusiastic reportage on the action intellectuals with some rather sobering historical reflections. Reaching all the way to Thucydides some twenty-four hundred years ago, he recalled for us Thucydides' warning of the dangers posed by intellectuals in power: "It was, he said, frequently a misfortune to have very brilliant men in charge of affairs, they expect too much of ordinary men." And such men, who "seek to impose logic on a people too fast," before they are ready, are dangerous to the people's welfare and security.[62]

NOTES

1. Cited in Steve J. Heims, *John von Neumann and Norbert Wiener: From Mathematics to the Technologies of Life and Death* (Cambridge: The MIT Press, 1980), p. 371.

2. See Col. Harry G. Summer, Jr., "What Is War?" *Harpers*, May 1984, pp. 75–8.

3. Quoted in Lawrence Freedman, *The Evolution of Nuclear Strategy* (New York: St. Martin's Press, 1981), p. 48.

4. Michael Howard, *The Causes of War* (London: Unwin Paperbacks, 1983).

5. In H. Rothfels, "Clausewitz," in *Makers of Modern Strategy*, ed. Edward Mead Earle (New York: Atheneum, 1966), pp. 99–100.

6. Engels, cited in Earle, ed. *Makers of Modern Strategy*, p. 158.

7. Cited in Robert Osgood, *Limited War: The Challenge to American Strategy* (Chicago: Chicago University Press, 1957), p. 33.

8. In I. Deutscher, *The Prophet Armed* (Oxford: Oxford University Press, 1976), p. 484.

9. N. N. Inozemtsev, "Why U.S. Area Studies Are Vital," cited in *The Current Digest of the Soviet Press*, vol. XXII, no.11, April 14, 1970. See also, William and Harriet Scott, "The Social Science Institutes of the Soviet Academy of Sciences," *Air Force Magazine*, March 1980; and Carl G. Jacobsen, "Soviet Think Tanks," *Soviet Armed Forces Review Annual*, Fall 1977, pp. 140–52.

10. Leo Oppenheim, "The Positions of the Intellectual in Mesopotamian Society," in *Daedalus*, Spring 1975, pp. 37–57.

11. Robert Nisbet, *Twilight of Authority* (New York: Oxford University Press, 1975), p. 176.

12. Ibid., pp. 154–74.

13. Ibid.

14. Raymond Aron, *The Opium of the Intellectuals* (New York: W. W. Norton & Co., 1957).

15. Nisbet, *Twilight of Authority*, pp. 205–06.

16. Edward Shils, *The Intellectuals and the Powers* (Chicago: The University of Chicago Press, 1972), p. 172.

17. W. G. Runciman, ed., *Weber: Selections in Translation* (Cambridge: Cambridge University Press, 1978), p. 348.

18. Nisbet, *Twilight of Authority*, p. 177.

19. Alexis de Tocqueville, *The Old Regime and the French Revolution*, cited in George DeHuszar, *The Intellectuals: A Controversial Portrait* (New York: Doubleday & Co., 1955), pp. 11–18.

20. See Richard Hofstadter, *Anti-Intellectualism in American Life* (New York: Vintage Books, 1962), p. 219.

21. In Theodore H. White, "Action Intellectuals," *Life*, June 9, 16, 23, 1967.

22. "The Problem Solvers of National Security," *Business Week*, July 13, 1963.

23. Wesley W. Posvar, "The Impact of National Expertise on the National Security Policy of the United States," *Public Policy* XIII (1964), pp. 36–64.

24. White, "Action Intellectuals."

25. Ibid.

26. Ibid., p. 23.

27. de Tocqueville, *The Old Regime.*

28. Bernard Brodie, ed., *The Absolute Weapon* (New York: Harcourt Brace, 1946), p. 76.

29. The most influential and best known defense intellectuals were associated with the Rand Corporation, the Institute for Defense Analysis, the Hudson Institute, and the research institutes associated with Harvard University, Columbia University, and MIT, among others, as well as with the bureaucracies of the Pentagon and the Central Intelligence Agency. For discussion of the various relevant institutions and staff members, see Arthur Herzog, *The War–Peace Establishment* (New York: Harper & Row, 1965), particularly chapter II, "The Analysts"; Paul Dickson, *Think Tanks* (New York: Atheneum, 1972), chapter VII; Saul Friedman, "The Rand Corporation and Our Policy Makers," *The Atlantic Monthly* September 1963, pp. 61–8; and Bruce L. Smith, *The Rand Corporation* (Cambridge: Harvard University Press, 1966).

30. Freedman, *The Evolution of Nuclear Strategy,* chap. 12, "The Formal Strategists."

31. Max Lerner, "Editor's Introduction," in *The Portable Veblen* (New York: Viking, 1948).

32. Ibid., p. 269.

33. Ibid., p. 232.

34. Benjamin Lambeth, "Contemporary Soviet Military Policy," in *The Soviet Calculus of Nuclear War,* ed. Roman Kolkowicz and Ellen Mickiewicz (Lexington, Ken.: Lexington Books, 1986), p. 35.

35. Ibid.

36. Ibid., p. 7.

37. Marshal of the Soviet Union V. D. Sokolovskii, *Soviet Military Strategy,* ed. H. Dinerstein, L. Goure, and T. Wolfe (Englewood Cliffs, N.J.: Prentice-Hall, 1963).

38. Cited in R. Kolkowicz, "The Soviet Union: Elusive Adversary," in *The Soviet Calculus of Nuclear War,* p. 6.

39. Ibid., p. 7.

40. L. Feokistov, "Atomic Bomb As It Is," in *Peace and Disarmament: Academic Studies* (Moscow: Progress Publishers, 1984), p. 76.

41. In Kolkowicz, "The Soviet Union: Elusive Adversary," p. 7.

42. Ibid.

43. Ibid., p. 8.

44. Eugeny Velikhov, "Averting a Nuclear Catastrophe," in *Peace and Disarmament,* p. 53.

45. See the chapters in this volume by Colin Gray (#5) and Benjamin Lambeth (#7).

46. See the chapter in this volume by Ken Booth (#9).

47. See the chapters in this volume by Robert Jervis (#2), George Quester (#3), Lawrence Freedman (#4), and Joseph Nye (#8).

48. See the chapter in this volume by Arthur Stein (#6).

49. Secretary of Defense Caspar W. Weinberger, "U.S. Defense Strategy," in *Foreign Affairs,* Spring 1986, p. 676.

50. See Fred C. Ikle, "Nuclear Strategy: Can There Be a Happy Ending?" *Foreign Affairs*, Spring 1985, pp. 810–11.

51. See Gary Taubes and Glenn Garelik, "How Good Is Soviet Science?" *Discovery*, August 1986, vol. 7, no. 8.

52. See General Brent C. Scowcroft's and others' negative views concerning arms control proposals for deep weapons reductions that "could increase the vulnerability of American nuclear forces to Soviet attack, and would leave us with a force posture that tends to be less stable than it is now" in Michael R. Gordon, "Experts Criticize Reagan's Proposal for Weapons Cuts," *New York Times*, Sunday, July 13, 1986, p. 1.

53. Sir Solly Zuckerman, *Scientists and War: The Impact of Science on Military and Civil Affairs* (London: Hamish Hamilton, 1956), p. 63.

54. Colin Gray, "What Rand Hath Wrought," *Foreign Policy*, no. 4 (Fall 1971).

55. Freedman, *The Evolution of Nuclear Strategy*, p. 185.

56. Ibid., p. 184.

57. Cited in White, "Action Intellectuals."

58. Karl Mannheim, *Ideology and Utopia* (New York: Harcourt, Brace, 1936), p. xiv.

59. Paul Hollander, *Political Pilgrims* (New York: Harper, 1981), p. 41.

60. F. Ikle in "Can Deterrence Last Out the Century," *Foreign Affairs*, January 1973, p. 14.

61. Fritz W. Ermarth, "Contrasts in American and Soviet Strategic Thought," in *Soviet Military Thinking*, ed. D. Leebaert (London: George Allen & Unwin, 1981), p. 64.

62. In White, "Action Intellectuals."

CHAPTER 2

Strategic Theory: What's New and What's True

ROBERT JERVIS

CONTINUITY IN STRATEGIC THOUGHT—BRODIE AND BORDEN

Bernard Brodie was one of the first people to think seriously about the impact of nuclear weapons on world politics. It thus is useful to ask how more recent arguments relate to his. We want to know what is new, what is true, and what the relationships are between these two categories. It seems to me that much of what Brodie wrote forty years ago contains a great deal of truth and that much of the effort to develop new ideas has been misguided.[1]

The centerpiece of Brodie's position was the claim that nuclear weapons revolutionized both military strategy and the relationships between force and foreign policy.[2] In the past, the ability to destroy others' armed forces and the ability to protect oneself were tightly linked. In the past also, weapons of a particular kind usually fought their opposite numbers, that is, tanks fought tanks, infantry fought infantry, and battleships fought battleships. Nuclear weapons changed this. They are so powerful and hard to defend against that one side's nuclear forces would not fight those of the other side. Rather, the target for atomic bombs would be what each side valued most.

This difference is enormous in its consequences—no one could win a major nuclear war. To use later terminology, deterrence by punishment replaces deterrence by denial.[3] States would be dissuaded from going to war not by the belief that they could not defeat the other side's military forces, but by the knowledge that they would be destroyed in the process. Crucial here is the potentially mutual nature of deterrence—if both sides have second-strike capability, each can destroy the other, and neither can defend itself. It follows that, unlike

The author would like to thank Joseph Nye, George Quester, and John Ruggie for their comments.

the situation in the prenuclear era, absolute power is more important than relative power. When armies fight armies, knowing the absolute size of the forces on either side tells one nothing because what matters is the relative capability of the two forces that will clash. But with nuclear weapons, what is crucial is each side's ability to cripple the other, and this capability is measured by the match between the country's forces and the targets it seeks to destroy, not between the two sides' forces.

Brodie's views were certainly incorrect when he first enunciated them because nuclear weapons were scarce and, by today's standards, small. It would be a number of years before the United States could destroy the Soviet Union with nuclear weapons. Thus, the first careful study of the utility of atomic bombs—the Harmon Report of 1949—concluded that the American bombs were too few, too small, and carried by planes that were too vulnerable to allow the Air Force to fight the war alone. (Perhaps not surprisingly, the Secretary of Defense withheld the report from President Truman.[4]) It took the H-Bomb to bring about the world Brodie had foreseen. Interestingly enough, Brodie's initial reaction to this development was to stress the need for the ability to fight limited wars on the grounds that the immense destructive power that would soon be available to the superpowers made all-out war impossible, an issue that is a main focus of debate today.

Of course, Brodie was not the only pioneer. Not only were other people working out similar ideas, but William Liscum Borden also made counterarguments strikingly similar to those of current American nuclear doctrine. In his little-known *There Will Be No Time*, Borden saw nuclear weapons not as revolutionizing strategy, but as reinforcing many of its traditional elements.[5] Because the vulnerability of both sides' cities would lead each side to spare these targets, atomic bombs would be used against military installations. Capability therefore still had to be measured in relative terms, and, just as in the past, preventing the adversary from gaining a military advantage was the essential point of strategy. Although this view did not deny the obvious fact of the enormous explosive power of nuclear weapons, it saw them as not qualitatively different from those with which we were familiar. There is no reason to think that Borden would have disagreed with Nitze's assessment thirty-five years later:

> It is a copybook principle in strategy that, in actual war, advantage tends to go to the side in a better position to raise the stakes by expanding the scope, duration or destructive intensity of the conflict. By the same token, at junctures of high conflict short of war, the

> side better able to cope with the potential consequences of raising the stakes has the advantage. The other side is the one under greater pressure to scramble for a peaceful way out. To have the advantage at the utmost level of violence helps at every lesser level.[6]

As this quotation indicates, the most important question is also the most obvious one: Can there be a meaningful victory in nuclear war? For Borden, Nitze, and many others, the answer is yes (or at least the Russians might believe the answer to be yes). Destruction would be enormous, but one side could end up ahead, and this possibility could influence peacetime bargaining. It follows that American military strategy must resemble the general outlines of prenuclear strategy. The United States must be prepared to fight a nuclear war or, rather, to fight the variety of nuclear wars that might occur. Of course, this does not mean the United States should seek such wars. The purpose of the capability is deterrence. But deterrence works through having the potential to gain victory if need be. Thus, America requires the capability if not to prevail in a nuclear war, then at least to be able to terminate the war on favorable terms.

For Brodie and many others, victory is not possible. Or, to put it more precisely, military victory in a nuclear war is not possible. Crises and even limited wars could be terminated in ways highly advantageous to one side and disadvantageous to the other, but as long as both sides remain vulnerable, the balance of military power is not decisive because it cannot protect either side from destruction.[7]

Of course, not all the basic ideas were on the table by the end of 1946. Without trying to give a history of the development of strategic thought,[8] I think the development of three concepts in the 1950s fleshed out the earlier ideas. Wohlstetter's distinction between first- and second-strike capability, Snyder's clarification of the problem of extended deterrence, and Schelling's discussion of the reciprocal fear of surprise attack—what is now called crisis stability—were crucial in providing the conceptual tools for our current strategic thinking. These developments added sophistication and nuance to the positions associated with Brodie and Borden but left the basic arguments unchanged.

It is sometimes argued that these ideas can no longer explain national behavior or shape policy because circumstances have changed from those that prevailed when they were developed. As Secretary Weinberger asked: "Should ideas formulated in an era of American military predominance apply with equal validity in an era of parity?"[9] A negative answer seems self-evident, but the question is misleading on two counts. First, although the 1950s and 1960s may now appear to be eras in which the Soviet military challenge was relatively slight, this was not the contemporary perception. Second, and more important, although parity was still some years away, the strategic ideas were developed with this configuration in mind. Both Brodie and Borden were writing not for the present, but for the inevitable future when both sides would be well armed.

Importance of Concepts

As the current strategic debate shows, arguments become garbled unless we maintain the conceptual clarity provided by the distinctions between deterrence by denial and deterrence by punishment. Current American strategy stresses the importance of denying the adversary any advantage from starting a conflict at any level of violence. At first glance, this seems a straightforward argument for deterrence by denial that although perhaps misguided, is at least coherent. But a closer examination of the rationales provided by the position reveals that this is not the case. Here are a sample of official explications of this theme: The United States needs to have the forces necessary "to deny an enemy his objectives rather than to face him solely with the prospect of Pyrrhic victories." "For deterrence to operate successfully, our potential adversaries must be convinced that we possess sufficient military force so that if they were to start a course of action which could lead to war, they would be frustrated in their efforts to achieve their objective *or* suffer so much damage that they would gain nothing by their action." "Our forces must be in a position to deny any meaningful objective to the Soviets *and* impose awesome costs in the process." "Our basic strategy requires us to be able to inflict such damage on a potential adversary, that regardless of the circumstances, the prospect of that damage will preclude his attack on the United States, our allies, or our vital interests." The United States needs to leave the enemy "no illusion of making any gain without offsetting losses."[10]

As the italicized words indicate, sometimes the call is for deterrence by denial, sometimes for deterrence by punishment, and sometimes either one or the other is seen as sufficient. The whole analysis then makes no sense because it does not distinguish between the ways force brings pressure to bear. By neglecting the clarifying tools that early work in strategic analysis provided, the current statements both fail as guidelines to coherent policy and inadvertently pay tribute to the power of the older ideas.

Two points need to be stressed. First, the relevant distinction is between deterrence by denial and deterrence by punishment, not between counterforce and countervalue. Second, although this distinction clarifies many issues and policies, some targets partake of both categories. Turning first to the nature of the distinction, the use or threat of force exerts pressure either by making it impossible for the other side to reach its objectives or by exacting an unacceptable price for doing so. The threat to destroy civilian targets usually fits into the latter category and the threat to destroy military targets into the former, but this connection is not rigid and invariant. Thus, in a long conventional conflict, destroying civilian targets such as factories could prevent the other side from carrying on the war. The opposite side of this coin is that destroying military forces indirectly involved in the war does more to punish the other side than to affect the course of battle. For example, in a war limited to Europe, the American threat to destroy Soviet forces in the Far East would be much more punishment than denial.

Just as deterrence by punishment is too often identified with purely countervalue targeting, so it is often inappropriately implied that punishment must be meted out quickly and suddenly. That is, people tend to associate a strategy based on the ability to destroy what the other side values with a *spasm* attack that would carry out all the destruction at once. Although there are good reasons to doubt that any nuclear war could be kept limited, it is a mistake to equate the belief in the efficacy of punishment with the belief that targets must be destroyed in the first hours of a war. Indeed, it may be much easier to destroy some cities and not others than it would be to wage a limited war of denial. As Schelling has noted, it is just because what the other side values is so important to it that destroying all such targets would make no sense; bargaining leverage comes from the existence of targets that have *not* been struck.[11]

The clarity of the theoretical distinction between deterrence by denial and deterrence by punishment does not mean that all targets fit easily into one category or the other. Some obviously fit both. The most important example is the set of targets crucial to the Russians' control of their country and empire (e.g., the party structure, internal security forces, and conventional forces in East Europe). To strike these would destroy what the Soviets value and deny them the ability to conquer and control West Europe—and perhaps East Europe as well. This does not mean that a strategy based on holding such targets at risk would be particularly efficacious; problems of practicality, credibility, alternatives, and costs would have to be considered. But it does remind us that not all targets and strategies neatly fit our categories. Enough do, however, so that the distinction is extremely helpful for clarifying thinking as to why threatening or destroying certain targets should be expected to deter the other side or terminate the war.

Incentives to Innovate

In the realm of ideas, there are great incentives to innovate. Credit goes to people who say something original and interesting. It helps to be right, but usually this is not necessary. Thus, academics bemoan "dead" fields in which there are few new ideas: How can one write a book if everything significant has been said? Further pressures to come up with something new are provided by the dissatisfaction with the state of current arguments and the belief that progress, if not inevitable, should certainly be possible. This chapter later discusses some reasons we might expect intellectual improvement and why, in fact, progress has been relatively slight.

Policy makers also have incentives to innovate. The military often needs new rationales for increased budgets, although the fact that the armed services want to continue to perform established missions provides an important brake on the development of new ideas. Those people responsible for developing war plans are likely to feel pressures for new ideas stemming from sources both bureaucratic and psychological. Persons charged with trying to think of what the United States should do in the event of war will have to consider the widest range of possibilities and, unless they are fooling themselves, will never be satisfied by the responses that are available. Therefore, they will be driven to develop ever more elaborate, complex, and subtle plans. At the highest levels, the need of each incoming administration to distinguish its policy from that of its predecessor provides incentives to announce, if not to construct, a policy that is new and different. Thus, Secretary Weinberger began a recent statement on American defense policy by saying: "When the Reagan Administration entered office in 1981, a lengthy debate could have been held on the question: 'Which was in worse shape—US military hardware or US strategic concepts?' "[12] Developing modern ideas, he went on to argue, was as important as developing modern weapons.

The problem, however, is that whereas irresponsible academics can afford to be original at the expense of being correct, statesmen cannot. What is most important is that the policy be an appropriate one. Administration officials have called for a "renaissance in strategic thinking" without much consideration of the fact that originality is not automatically good.[13] We are better off following old, good ideas than developing new, bad ones.

The distinction between what is new and what is true produces a fourfold table as in Figure 2.1. Figure 2.1 provides a useful way to organize many of the ideas in nuclear strategy.

	Valid	Invalid
Old	Nuclear revolution Crisis stability	Deterrence by denial Counterforce
New	Deterrence and cooperation Psychological factors	Complex scenarios Overelaborate war plans and target lists

Figure 2.1

OLD, GOOD IDEAS

The most important old, good ideas have already been mentioned. Primary is the concept of the nuclear revolution with the concomitant implication that many prenuclear ideas are now inappropriate. Perhaps most central is the claim, now bitterly contested, that the state does not need the ability to take meaningful military action in order to deter depredations. Because any extreme crisis and, even more, any use of force between the superpowers creates a chance of mutual destruction, military advantage loses most of its traditional meaning. Gaining some territory or destroying more of the other side's warheads will not do an attacking state any good if the end result is all-out war. Furthermore, escalation to total violence not only can occur as the result of an explicit decision to reach that outcome, but also can be produced by a series of smaller steps, each one seeming rational, by the unintended consequences of emotional outbursts, or by unauthorized lower-level military decisions. Schelling states the point very well:

> The idea, expressed by some writers, that . . . a country cannot plausibly threaten to engage in a general war over anything but a mortal assault on itself unless it has an appreciable capacity to blunt the other side's attack, seems to depend on the clean-cut notion that war results—or is expected to result—only from a deliberate yes–no decision. But if war tends to result from a *process*, a dynamic process in which both sides get more and more deeply involved,

> more and more expectant, more and more concerned not to be a slow second in case the war starts, it is not a "credible first strike" that one threatens, but just plain war. The Soviet Union can indeed threaten us with war: they can even threaten us with a war that *we* eventually start, by threatening to get involved with us in a process that blows up into war.[14]

The result is that the United States does not need to meet all Soviet threats on their own terms, as Brodie realized. Tripwires can be very effective. As long as using force to change the status quo sets in motion forces that are hard to control, taking advantage of local military imbalances is extremely dangerous. Thus, although Brodie's initial reaction to the development of the hydrogen bomb was to argue that the United States had to prepare to fight limited wars, by the early 1960s he felt that the Kennedy administration was pushing this idea much too far, arguing that the West did not need the ability to defend Europe with conventional weapons. The pattern he located in past confrontations would continue to hold:

> The essential question was: How much did the strictly local forces affect the Russians' willingness to risk open hostilities with us—or, for that matter, our willingness to get into such hostilities with them? And the answer clearly was, little or none—unless both sides had been *completely convinced* that a quite considerable shooting war could develop on the spot without its substantially widening and incurring extremely grave risk of nuclear weapons being introduced.[15]

Because neither side can expect to win a nuclear war, such a conflict is extremely unlikely. Because lower level clashes can escalate, they too are likely to be deterred by the fear of destruction. Mutual vulnerability then means that the superpowers are both highly secure and insecure at the same time—secure because the chance of war or menace to their vital interests is low; insecure because if war occurs, it could bring annihilation. Indeed, it is this chance that protects the states.

OLD, BAD IDEAS

Most ideas that are old but bad start from the belief that mutual vulnerability either can be overcome or need not be of central importance. As Brodie noted in *Strategy in the Missile Age*, the fact that the Soviet Union could destroy the United States has led to a number of attempts to escape from this situation and reestablish traditional routes to security.[16] Perhaps the most powerful and lasting ideas of this kind are the politically opposed but intellectually linked desires to take our cities out of hostage and to abolish nuclear weapons. Both impulses arise from feelings about what is necessary, not analyses about what is possible. The current ABM program is not derived from a considered judgment that the problem can be solved. Instead, it is driven by the feeling that the alternative to defense is unacceptable. Thus, the emotion behind Reagan's position comes through in his extemporaneous remark: "To look down an endless future with both of us sitting here with these horrible missiles aimed at each other, and the only preventing a holocaust is just so long as no one pulls this trigger, this is unthinkable."[17]

At the other end of the political spectrum, but driven by the same fear, is the drive to abolish nuclear weapons. The popularity of Jonathan Schell's recent books shows the strength of feeling here, and it is interesting that in the second of them he argues that ABM systems will make abolition possible.[18] Indeed, the United States remains committed to the idea of eventually doing away with all nuclear weapons, although there are few indications that decision makers have thought seriously about either the desirability or the feasibility of such a drastic change.[19]

A second idea that is both old and bad is what Hans Morgenthau called "conventionalization."[20] Many analyses of nuclear weapons use the intellectual framework that was appropriate in an earlier era, when maintaining a military advantage could protect one's values. Defense by denial was possible; pushing back the other side's army could both achieve your own goals and protect your citizens. This was Borden's point of view, and it informs current American nuclear policy. Even if nuclear weapons continue to exist and cities cannot be removed from hostage, partial security can still be reached through the military ability to prevent the other side from reaching its goals in third areas, for example, being able to defend Europe and the Persian Gulf against nonnuclear attacks. Because nuclear weapons cancel each other out, conventional defense by denial is necessary and possible. In the words of Robert McNamara, the "sole purpose" of each side's strategic forces "is to deter the other side's first use of its strategic forces."[21] But this position would make sense only if escalation were extremely unlikely. If lower levels of violence were hermetically sealed off from higher ones, then mutual vulnerability would mean that only limited force could be used and that these exercises of military power would operate as they did in the pre-nuclear era. In fact, however, statesmen cannot dismiss the danger that what starts out as controlled violence could lead to all-out war. Thus, conventionalized perspectives cannot explain past Soviet–

American interactions or be a guide to effective policy, although they do help explain many aspects of American war planning.[22]

The third idea that is old and bad is an error of omission rather than commission. This is the ignoring of the security dilemma—the fact that under many circumstances an increase in one state's security will automatically and inadvertently decrease that of others. The idea of the security dilemma is very old—Thucydides saw this dynamic as the root cause of the conflict between Athens and Sparta. It received renewed attention in the late 1940s and early 1950s, but not by people who were specializing in nuclear weapons.[23] Partly because of the natural self-righteousness of the first years of the Cold War, statesmen and analysts gave little consideration to the impact of United States policy on legitimate Soviet security interests. For example, the Acheson–Lilienthal program, let alone the Baruch Plan that the United States finally proposed, was a sufficient threat to Soviet security so that almost any Russian leader, even if he were neither paranoid nor aggressive, would have rejected it. But few American experts realized this. Although they expected the Russians to reject the proposal, the basis for this prediction was their belief that the Russians were aggressive, not that the proposal was unreasonable. Indeed, the Russian reaction was taken as confirmation of this dark view of their motives.

In the succeeding years, much United States policy and many aspects of strategic analysis fit the same pattern. In the early 1960s, even after it was known that the missile gap did not exist, few people worried about the effect of the American buildup on Soviet security or the extent to which the policy might lead the Russians to view the Americans' expressed desire for equitable arms control arrangements as hypocrisy. (The Russians seem to have had a similar inability to appreciate the unintended consequences of their own actions.)

NEW, GOOD IDEAS

Understanding the Security Dilemma

Although the bulk of the ideas, good and bad, that dominate strategic thinking date from the first fifteen years of the nuclear era, some are fairly new. The first new and good idea is the opposite side of the coin noted previously. We are now much more aware of problems created by the security dilemma. This has important implications for both policy and scholarship, although progress is more marked in the latter area. This awareness leads to a better appreciation of the possibility that two countries, both seeking security, can become involved in conflict and even wars with each other. States may inadvertently create spirals of tension and hostility. Unfortunately, it is extremely difficult, even after the fact, to determine each side's motivation and to judge whether more conciliatory policies would have dampened the conflict. For example, scholars still debate whether World War I could have been avoided, and if so, whether firmer or more conciliatory policies would have been required.[24]

The problem is twofold: it is difficult to determine the other side's intentions, and trying to deter a state that is primarily seeking security is likely to create unnecessary conflict. Two implications follow. First, deterrence theory's prescriptions are conditional. Deterrence should be stressed only in certain situations; applied inappropriately, it can heighten the security dilemma. Second, because it is often impossible to be certain of the other side's intentions, statesmen should try to design policies and postures that will protect their security while decreasing the other's security as little as possible. An understanding of the security dilemma can lead statesmen to seek to ameliorate it.

There are, of course, severe limits on what is possible here. Even when technology is most propitious, protecting oneself menaces others at least to some extent. And although there is some choice as to which technology to employ, freedom of action is far from complete. For example, it would be difficult for the United States and the Soviet Union to return to a world of invulnerable ICBMs, even if they desired to. Furthermore, technology is not the only factor that intervenes between an understanding of the security dilemma and the designing of relatively nonthreatening policies. States' security interests often are extensive and may require offensive capability to meet them. Thus, an American posture that requires the credible threat to initiate nuclear war in a response to a Soviet conventional attack on Europe or the Persian Gulf must appear to the Soviets as a posture that might be used to change the status quo to their disadvantage.

But if a purely defensive and nonthreatening posture is never possible for a major power, not all possible responses are equally threatening to the other side's security. There is at least some domain of choice, and statesmen can act to minimize the unintended consequences of their quest for security. For example, making one's missiles less vulnerable or mobile threatens the other side less than does increasing their numbers. Similarly, NATO could increase its security without menacing that of the Warsaw Pact countries if it were to fortify the East–West border, especially in Germany. (Unfortunately, the realities of German domestic politics make this almost impossible.) By contrast, whatever its other merits, a NATO strategy based on a counterattack into East Europe or on strikes deep behind Warsaw Pact lines would have the cost of decreasing Soviet security, just as the Soviet offensive strategy menaces the West and leads to the inference that Soviet intentions are malign.[25]

Three implications follow; the first is hotly disputed and the other two have often been lost sight of in recent years. First, the state may not only seek to make its own forces invulnerable but may also want its adversary's forces to be invulnerable. Decreasing the other side's security may make the state less rather than more secure.

Second, if mutual vulnerability is what is most dangerous, then arms control should concentrate on the characteristics of weapons, not on the numbers of them.[26] Increasing one or both sides' arsenals may not be destabilizing; freezing or decreasing levels may not make either side more secure. What is important is whether the countries' weapons are vulnerable or can destroy the other side's strategic forces.

Third, stability in a crisis can be enhanced if forces are designed so that the measures each side would take, if it feared war, would increase its ability to retaliate without simultaneously increasing its ability to launch a first strike. Thus, to the extent that mobilization involves making larger numbers of weapons ready to fire, the state not only increases the number of weapons that are likely to survive an adversary's strike, but also gains a greater ability to launch a first strike. Similarly, many aspects of putting the command, control, and communication (C^3) system on alert would be necessary for launching a first strike as well as for striking back.

In principle, some measures would be purely defensive, for example, moving mobile missiles from known bases to hidden ones. But these are rare and, furthermore, in the heat of a crisis, are likely to be perceived by the other side as evidence of a likely first strike, even if they do not logically put the other side in a position to start the war. Nevertheless, not all measures to protect oneself threaten—and are seen as threatening—the adversary to the same degree, and crisis stability is enhanced to the extent that we can design postures and weapons systems that look different when they are being manipulated in preparation for a first strike than when they are being made secure against a possible attack.

Even if this can be done, however, stability is threatened if the state's retaliatory structure could be crippled if it were struck before it went on alert. Indeed, the greater the gap between a state's ability to retaliate under normal circumstances and its ability to do so when it is on alert, the greater the incentives for the other side to attack before the state can move into the latter posture. The temptation to preempt in times of calm would be very slight; under even the best conceivable circumstances, the chance of wiping out the other side's retaliatory ability is low and the costs of a less than completely successful preemptive strike are enormous. But the incentives would be different in the unlikely but dangerous circumstance that one side believed a serious crisis was inevitable in the near future. It would then see a "window of opportunity," which would be closed as soon as the other side saw the danger and put its forces on alert. There would then be some danger if the United States developed systems of weapons and C^3 that, although invulnerable when on alert, might be successfully attacked when they were not. Thus, it would be best if the systems could be invulnerable all the time, although the costs of such systems, both in terms of money and in terms of being on a hair trigger, might be excessive.

Possibilities of Cooperation

An appreciation of the security dilemma has led to an analysis of security from the perspective of the possibilities of increasing cooperation. Even if each side's preferred choice is to take advantage of the other, mutual cooperation is possible if big transactions are broken up into small, repeated ones, if defensive postures are stronger than and different from offensive ones, and if, in any single instance, exploiting the other does not produce overwhelming gains and being exploited by the other does not inflict overwhelming losses.[27] A related approach has been developed by scholars working on the question of international *regimes*. They argue that, even in the absence of formal international institutions with powers of enforcement, the development of common principles, norms, and rules can facilitate cooperation in situations characterized by a mixture of common and conflicting interests that closely resemble those described by the security dilemma.[28]

There is, thus, a convergence of two fields of study that usually ignore and implicitly contradict each other—the field of security policy and the field of international organization and the coordination of international political economic activities. The former has tended to stress conflict, the latter cooperation; the former has taken a realist perspective, the latter has embodied elements of legalism and idealism. But studies of security that stress the security dilemma and the concomitant concern with the possibility of unnecessary conflict have much in common with analyses of international political economy, which deal with the dilemmas created by the fact that economic cooperation is not automatic even when the actors prefer this outcome to one in which they all pursue highly competitive policies. Students of security see that states must avoid threatening others unnecessarily; students of regimes realize that cooperation can only be maintained if each state is able to retaliate against others' transgressions.

In other words, both fields are concerned with similar problems of how states can determine whether the other side is living up to implicit or explicit agreements, maintain the credibility of threats and promises, and protect their interests without setting off a cycle of mutually destructive responses. This does not mean that we can transpose all ideas of deterrence onto such areas as the economic relations among the advanced industrial countries or that we can adapt what we know about the workings of GATT to an understanding of Soviet–American arms control. Some of the problems are the same, however; some kinds of explanations apply in both arenas, and there are parallels in the kinds of policies that will be effective.[29]

In both security and political economy, the hope to exploit the other and the symmetrical fear of being exploited can drive states into conflict, even if they know that mutual cooperation would be better than a high level of strife. In this situation, the latter outcome can be avoided if each side can convince the other that it will cooperate *if and only if* the other does so. Making either threats or promises credible is difficult enough; doing both simultaneously is especially demanding. For example, President Carter probably succeeded in convincing the Russians that he would cooperate, but he also tempted them to exploit him. Reagan convinced them that he would respond to their lack of cooperation by a similar stance of his own, but he left them with grave doubts as to whether he would also reciprocate concessions.

Students of security policy recognize that the state needs to show that it does not have to cooperate and will react strongly to the other's transgressions. However, these students have not paid sufficient attention to the other side of this coin—the need to convince the other that the state will cooperate if the other does. If the other side believes that the state will harm it no matter what, it will be impossible to bring about any common interest. Therefore, states must provide their adversaries with reassurances as well as threats. Indeed, as discussed at the end of this chapter, the failure to do this is the most likely way a world war would start. Furthermore, I would propose the hypothesis that American foreign policy toward the Soviet Union has more often suffered from the difficulty of making credible promises than the difficulty of threatening. Do the Soviets believe our defense programs will be restrained if theirs are? Do they think the West will abstain from interventions in the Third World if they do likewise? Do they think renewed detente would bring credits rather than a continuation of the Jackson–Vanik and Stevenson amendments? One can, of course, argue that these bargains would be unacceptable to the Soviet Union or should be unacceptable to the United States, but it is hard to maintain that the Soviets have reason to be confident that the Americans would match their cooperation.

To make one's behavior conditional on that of others requires the ability to determine what others are doing, in other words, intelligence and inspection. Although in the security field most of the discussion of inspection has been connected with arms control agreements, the need is present even without them; indeed, the United States developed satellite surveillance systems before there were any treaties to verify. With or without formal agreements, inspection can increase cooperation and mutual security both by alleviating each side's fear that the other is greatly increasing its arms and by making it unlikely that either could think that it could engage in a buildup without being detected.

The state not only wants to observe its adversary, but it also, under many conditions, wants its adversary to know what it is doing. This is the case when one state is willing to be restrained if the other is and believes that the adversary is more likely to keep its level of arms low if it believes the state is also following this policy.

Similarly, attempts to maintain a relatively open international economic system require that each state know the extent to which others are constructing protectionist barriers. In eras when the main barriers to trade are tariffs, the relevant information is readily available and unambiguous, thus facilitating cooperation. In recent years, however, nontariff barriers have become more important, and these are difficult to measure. Thus, the international economic realm may no longer be more transparent than the security arena. Indeed in the latter, strategic weapons are usually relatively easy to count, although the characteristics of the vital associated apparatus, like military doctrine and command and control arrangements, are usually much more opaque. Furthermore, general foreign policy behavior usually is hard to interpret. For example, as I write (May 1985), the newspapers carry stories about the foiled South African covert mission to Cabinda. It is clear that this was a violation of the agreement with Angola. But exactly how serious was the violation? Was the aim of the mission sabotage or merely to gather information, particularly information on whether Angola was abiding by the agreement? Was the South African action unprovoked or was it a response to previous violations by Angola? As the following discussion indicates, the difficulty in answering these questions and the tendency for each side to reach different conclusions about them complicate the attempts to maintain agreements.

Even if both sides are willing to cooperate, accurate information is only a necessary, and not a sufficient, condition for avoiding unnecessary conflict. The information must be received in time for the state to be able to respond. Thus, the notion of timely warning, developed in discussions of the Non-Proliferation Treaty, applies more broadly. Cooperation is much easier to establish when each side does not fear it will have suffered irreparable damage by the time it has detected and responded to the other's noncooperation. The prospects for cooperation are increased by breaking up one large transaction into a series of smaller ones.[30] Similarly, arrangements that lengthen the amount of time that would intervene between the state's commencement of noncooperation and the point at which harm would be inflicted on others can make it easier for states to cooperate. If states are confident that they will be able to detect and respond to cheating before much damage is done, they will be more willing to enter into agreements, feeling they have a safety net to protect them.

Of course, timely warning requires that the state be able to act on the basis of the information. In the best circumstances, these actions would have several characteristics. First, they would not harm or threaten states that were continuing to cooperate. Innocent bystanders would not be hurt, and only slight damage would be done to the other side if the state incorrectly believed that the other was not cooperating. Second, the actions of the state would not be so difficult or expensive to take that others would find its threat to be incredible. Students of security policy have been preoccupied with this requirement since 1945; students of cooperative arrangements in international economics are also becoming aware of its importance. Third, the actions or their effects should be reversible. That is, it should be clear to the other side that cooperation can be restored if it ceases trying to exploit the state. Fourth, it is often not enough that a state be able to protect itself; in order to deter the other side from transgressing, it may need to be able to inflict extra costs on the other that would more than nullify the other's gains from previous noncooperation. Of course, these requirements often are in tension with each other. For example, to the extent that it is cheap for a state to retaliate against the other's refusal to cooperate, it is also likely to be relatively tempting for that state to break the cooperative arrangement itself.

Even in domestic society, where centrally imposed sanctions are available, private responses are often both efficient and vital. Thus, in many instances, collaboration is enforced and maintained because the injured party can not only cease dealing with the other, but also spread the word that the other has acted dishonorably, making it less likely that third parties will be willing to enter into agreements with the other. Much of what is usually called trust is actually the availability of effective sanctions of this type, and communities in which such sanctions can be used will be more efficient than those that lack them, because they permit a wide range of bargains and an extensive division of labor.

Cooperation is then made possible because each side realizes that the other can halt its cooperation if provoked. Each must see that the other is not so devoid of alternatives that it has no choice but to go along. What is crucial, then, is not only the perception of the other's behavior, but also the attribution of that behavior to causes that both permit future cooperation and rule out the possibility of exploiting the other. Thus, the United States probably did not increase the chances of cooperation with the Soviet Union when it held down its spending on strategic arms in the decade from the late 1960s to the late 1970s. The Russians probably thought the restraint was not voluntary and so, on two counts, saw little reason to reciprocate (assuming they desired such an arrangement). First, they probably did not attribute the American behavior to a desire for better relations with the Soviet Union and so had no reason to expect that a regime of mutual restraint was possible. Second, if the American restraint was not responsive to Soviet behavior, there was no incentive for the Russians to be moderate. Rather, it was safe to seek unilateral advantage. When the Russians said that detente was irreversible, many Americans were reassured, but this position was disturbing because it implied that the United States had no choice but to maintain the relationship no matter what the Soviet Union did.[31] As deterrence theory has stressed, concessions viewed as stemming from weakness will not be reciprocated but will instead lead to increased pressure. Thus, experienced bargainers know that to seem too eager for an agreement is to invite the adversary to hold firm, if not to raise demands.

Equally destructive to the chance of mutual cooperation is the belief that the other side is already doing all it can to harm the state. Whereas the belief previously discussed removes the state's incentives to be moderate by leading it to conclude that it can make greater gains by exploitation, this belief removes the incentive by undermining the fear that shifting to a noncooperative policy will be met by an undesired response. Thus, Assistant Secretary of Defense Richard Perle buttressed his argument for abandoning SALT II by claiming that the Russians were doing as much as they could if there were no treaty.[32] Similarly, when President Reagan was asked whether the Russians might increase their arms spending in response to the American increases, he replied: "The Soviets—they're up at full pitch. I doubt if they could expand their military production beyond where it is right now."[33] He may believe that decreases in American spending would lead to Soviet restraint, but he does not fear that American increases would also be matched. Secretary of Defense Weinberger implicitly rules out both the possibility that our SDI program will provoke the Russians to proceed with similar systems and the possibility of restraining them by showing restraint when he claims that "the Soviet commitment to strategic defense is . . . unswerving."[34] (It should also be noted that people with beliefs like this usually also prefer mutual competition to mutual restraint, as Weinberger does in this instance.)

Similar inference processes operate with the predictions of whether the other will cooperate in the future. When the British considered their policy toward gas warfare, their continued restraint was made less likely by their estimate that, in the words of the official history of British intelligence, "in view of the measures [Germany] had taken to protect her civilians, the fear of retaliation was not likely to deter her from using gas if it was likely to be to her military advantage."[35] If the Allies had acted on this belief, they might have used gas when short-run calculations indicated there were gains from doing so because they would not have thought they could keep Germany restrained. Secretary of Defense Brown's quip about Soviet military programs yields a similar implication, although Brown did not act according to it: "When we build, they build; when we don't build, they build."

The need to make threats and promises conditional is encapsulated in the strategy of tit-for-tat. The state responds to the other's cooperative moves with cooperation of its own and reacts to hostility with hostility. If both sides prefer mutual cooperation to mutual hostility, if they understand each other's strategy, and if they correctly interpret each other's behavior, the result should be sustained cooperation, even though each side's first choice would be to take advantage of the other. This argument has many fascinating applications and implications, as Robert Axelrod has so lucidly shown.[36] But the final condition listed above often is not met because states misinterpret each other's behavior. Or, to put it slightly differently, states and their adversaries often differ in their interpretations. Random errors and random differences are possible. That is, essentially by accident, a state will sometimes see the other as having cooperated, when in fact it did not, or as having tried to exploit the state when it actually cooperated. Axelrod has shown that a 1 percent error rate does not destroy the efficacy of tit-for-tat, but Downs and his colleagues have argued that 1 percent is unrealistically low and have demonstrated that even a slightly higher error rate undermines the strategy, because so much noise is injected into the communications channels.[37]

Even more important than random errors are systematic biases. States tend to overestimate others' hostility and underestimate the harm they are doing to others.[38] It will appear to them that they are keeping their end of the bargain but the other side is not; that they are responding symmetrically whereas their adversary is indifferent at best; that the only way to protect their interests is to increase their own hostility. Perceptions like these probably accelerated the decline of Soviet–American detente in the 1970s and were at work in earlier cases as well—for example, the erosion of the Anglo–French entente in the 1840s.[39]

When states do not intend to harm others, they overlook the security dilemma and think that others understand that their intentions are benign. Almost universal is the belief that the other side understands that the state is willing to leave the other in peace if the other will only adopt a similar stance. Thus, the American decision makers failed to realize that in sending troops to the Yalu they decreased Chinese security. When the other responds with hostility, the state's inference compounds the problem. After the Chinese entry into the Korean War, Secretary of State Acheson argued that when American troops approached the Chinese border a few months earlier, "no possible shred of evidence could have existed in the minds of the Chinese Communist authorities about the [peaceful] intentions of the forces of the United Nations."[40] Since decision makers generally believe that the adversary understood that the state intended to cooperate, they conclude that the other's hostile response can only have one meaning—the other is highly aggressive, will take advantage of any opportunity to weaken the state, and will not reciprocate cooperation.

Difficulties in interpreting others' behavior and the differences in the two sides' perceptions also complicate a common prescription for maintaining cooperation. Both Axelrod's computer tournaments and some historical cases indicate that states should react quickly to the other's transgressions. As Axelrod notes, this was not a result he expected: "I came to this project believing one should be slow to anger. . . . It turns out that if one waits to respond to uncalled-for defections, there is a risk of sending the wrong signal. The longer defections are allowed to go unchallenged, the more likely it is that the other player will draw the conclusion that defection can pay. And the more strongly this pattern is established, the harder it will be to break it."[41] Alexander George's conclusion based on the decline of Soviet–American detente is similar:

> The danger of misperception and miscalculation is particularly serious when one superpower—typically the Soviet Union—intervenes cautiously and on a low level in a local situation, probing to test the other superpower's reaction and, when receiving no response, then decides to escalate its involvement. Failure to grasp that the opponent is conducting a carefully controlled probe and is engaged in a subtle asking of a question regarding the other side's intentions can easily abet miscalculation and set into motion escalation leading to confrontation.[42]

The problem, however, is that not only is an early response hard for a democracy, but also the quick determination of what is a defection is usually both difficult and contentious. For example, it is one thing to call for a prompt American response to Soviet violations of the SALT agreements and quite another to be certain when a violation has occurred. Although the passage of time, the gathering of more information, and the accumulation of more instances of the other's behavior do not always make the nature of what the other side is doing entirely clear, their absence usually makes governments hesitant to act because they realize that their estimates may be in error.

Quick responses are especially unlikely if the perceiving state expects long-run cooperation from the other, because this expectation—and the political stake in continued good relations—will color the interpretation of the other's behavior. Although a state will usually interpret others' behavior ungenerously, this is not true when it thinks that the other will cooperate. Furthermore, if it does react quickly, the outcome may be a spiral of conflict if the other believes that its behavior was legitimate or was a response to a previous defection by the state.[43]

The basic dilemma is unavoidable: a quick and firm reaction is appropriate in cases of unprovoked noncooperation. Without it, the other may conclude that it can make gains by frequent but minor cheating or, after its first defections go unpunished, may think that the state will acquiesce in unbalanced arrangements. But correctly perceiving the other's behavior is difficult, the room for differing interpretations is very great, and the appropriate match between provocation and response is fairly rare.[44]

Psychology and Security

As the previous paragraphs indicate, another fruitful new approach is the application of psychology to the problems of deterrence. Although the initial psychological criticisms of deterrence theory were often naïve about international politics in general and the Soviet Union in particular, they correctly stressed that deterrence theory was misguided in assuming that the only likely and dangerous problem was that status quo powers would fail to make clear their resolve to oppose aggression.[45] Two related problems are also troublesome: the exchange of threats between states that are actually willing to tolerate the status quo can lead to spirals of tension and conflict, and psychological processes explain why the credibility of deterrent threats may be greatly overestimated by the state making them or greatly underestimated by the state being threatened.

States are prone to underestimate the pressures others feel to change the status quo, pressures that can result in their misperceiving their environment and challenging commitments that are quite clear, credible, and potent.[46] Two steps are involved here. First, a statesman who is under grave international or domestic pressures may come to feel that a bellicose policy (either of supporting or challenging the status quo) is necessary. Second, in order to avoid the acute value trade-off, which would arise from an appreciation of the dangers inherent in the policy, the person will then come to see his stance as likely to succeed. Note that this approach does not, or at least not yet, incorporate the psychoanalytic factors Brodie felt were so important.[47] But it does begin to deal with the fact that deterrence involves human beings whose ways of thinking and feeling strongly affect their behavior.

New perspectives have also arisen out of the abandonment of the view that each state could adequately be characterized as not only a rational actor but also a unitary one. Analyses of bureaucratic politics have greatly enriched our knowledge of how American weapons are procured. Although we know a good deal less about the projected employment policy for strategic weapons, what we do know is consistent with the powerful influence of the way the military services have come to define their identities.[48]

What may be at least as important, but much less well understood, is the way in which international crises that involve the mobilization of extensive military forces need to be seen as the interaction of two unwieldy organizations. As John Steinbruner has argued, the military organizations are so large and complex that national leaders cannot maintain full control of or even fully comprehend them.[49] When systems are complex and tightly interconnected, they are prone to fail disastrously, even when the stimulus starting the process is fairly small, thus leading to Charles Perrow's unsettling concept of "normal accident," which our thinking about strategy does not readily comprehend.[50]

A high state of alert or the movement of military forces in a crisis would entail the implementation of unfamiliar operations that could lead to war through the operation of two linked mechanisms. First, the military forces on either or both sides could get out of control as authority is delegated down to operational levels and nervous commanders overreact—or perhaps just react—to the unprecedented behavior of the forces on the other side. Second, top decision makers may similarly misinterpret the other side's behavior as indicating that the other is about to attack. Since there has never been a situation in which both the United States and Russia went into a high-level nuclear alert (the Russians did not do so during the Cuban missile crisis, perhaps because they feared triggering American preemption), neither side would know how to interpret what the other was doing.

Of course, even if alerts were common, major problems of interpretation would remain, especially if the actions a state took to prepare for retaliation closely resembled those it would take to launch a first strike. However, the ambiguity and room for fear-driven perceptions are even greater when the decision maker is confronted with military activities he has never seen before. Furthermore, the very processes of entering a crisis mean that the other side's actions will be scrutinized especially carefully, and the result is likely to be that many suspicious happenings will be located. Even if the other's behavior has not changed, close scrutiny will usually lead the person to see things previously unnoticed. Thus, intelligence officers and decision makers may detect details of troop movements that, although routine, are not recognized as such because they had never been examined so carefully. The activities will then be seen as unusual and, given the context provided by the alert, as threatening. Furthermore, when these activities are examined, decision makers will give their attention to positive instances, that is, to evidence that is consistent with the other's plan to attack, and ignore or treat as insignificant the absence of behavior that, if present, would also indicate attack was likely. People generally fail to realize that their beliefs should be treated as hypotheses that can be undermined by the other side's inaction.

That it will be hard for each state to interpret the other's behavior in a crisis is perhaps less surprising than the fact that the decision makers will not understand their own state's behavior. When military forces are put on alert, they will execute all sorts of maneuvers of which the civilian leaders—and even the top generals—will be ignorant.[51] The actions taken in the name of the state may be more provocative than the leaders realize or desire. Furthermore, the leaders' misinformation means they will be especially slow to see the other side's actions as a reaction to what their state is doing. As a result, they will probably see the other's behavior as reflecting a settled decision to heighten the crisis, if not to attack.

Whether war would result would depend in part on whether decision makers believed that technology and tactics favored the offense or the defense, that is, whether they thought there was a significant advantage to striking first. Of course, beliefs and the objective situation can be very different, as we are reminded by July 1914, when both sides rushed to get in the first blow.[52] Decision makers again might overestimate the advantages of striking first. In fact, for all the discussion of the vulnerability of weapons systems and C^3, it is still extremely unlikely that even the best designed preemptive attack could deprive the other side of the ability to retaliate. Beliefs about the efficacy of a preemptive strike, however, might not reflect this reality.[53]

Strategy also is not entirely determined by objective factors, and it too strongly influences the existence and extent of an offensive advantage. Under current circumstances, if one side adopts a policy based on deterrence by punishment, the incentives to strike first are very slight because the retaliatory mission is relatively undemanding. On the other hand, if the state adopts a policy of deterring the other side by being able to maintain as much war-fighting ability as possible throughout a potentially prolonged armed conflict, then the gap between striking first and striking second is likely to be much greater. An adversary's attack that would be insufficient to prevent retaliation might be able to prevent the state from carrying out coordinated attacks against a wide range of military targets. Indeed, given the vulnerability of C^3 systems and land-based ICBMs, if one believes it is crucial to be able to stay ahead of the other side in the number of usable nuclear warheads, then striking first will surely be seen as quite advantageous. Of course, the costs of war are still likely to be seen as so enormous that a minimally rational statesman would prefer maintaining even an unsatisfactory peace to launching a strike. But if military advantage is deemed crucial, the latter alternative—bad as it would be seen—would still be so much better than receiving a first strike that crisis stability would be undermined. Thus, at least as important as technology are the decision makers' choices of strategy and beliefs about the offense–defense balance.

NEW, BAD IDEAS

In the final category are ideas that are both new and bad. Many of these ideas are embodied in current American nuclear strategy, although as I noted, the basic ideas of deterrence by denial can be found in Borden, and the concepts out of which this approach is built were developed during and appropriate to the prenuclear era. What is more recent is the almost baroque multiplication of options and targets. Although we still lack a complete history of war planning, it seems that a process of bootstrapping is involved.[54] A particular kind of target is put on the list, often at the initiative of civilians or because of a temporary surplus of warheads; the military accepts the logic and reports that since the number of targets in the new category is very large, the force is not sufficient to cover them. Thus, current strategy requires the United States to be able to hold at risk the communist control of the Soviet Union and Soviet conventional forces because, it is argued, these are the things the Russian leaders value. The number of warheads required is very large—much larger than the force that is available. Indeed, the number of military targets that we might want to destroy in conjunction with a war in Europe is almost limitless. The force requirements increase, furthermore, as the Russians increase the number of their installations.

Not only have we generated the requirement to destroy ever more targets, but we have also sought more selectivity, more control, and more precision. Starting from the perceived need to deny the other side a military advantage from any possible course of action, more thought leads to more imagined dangers, more limited options that could prove useful, and more ways in which coercion could be parried or employed. With these fears and intellectual tools, the obvious drive is to cover all contingencies, to think of all possible bargaining strategies. Greater and greater sophistication can then take us farther and farther down a blind and dangerous alley.

In the same way, many of our defense problems are of our own making. Assuming that the Russians draw straightforward inferences from our statements (a dubious assumption, to be sure), our claims about what is necessary for deterrence undercut the credibility of our threats by posing standards that are beyond our reach. For example, United States officials argue that deterrence requires the ability to fight a protracted nuclear war. But the vulnerabilities of C^3I, the tremendous stresses such a conflict would place on human beings and mechanical systems alike, and the unforeseeable frictions of an unprecedented kind of warfare mean that we cannot develop any reasonable confidence in our abilities to do so. A realistic appraisal of the situation, coupled with the unrealistic requirements for deterrence that we have set, could then undercut America's will.

On a more concrete level also, seemingly sophisticated thought creates artificial fears. For years, United States policy, supported by the analyses of many strategists, argued that Minuteman vulnerability presented grave dangers. Only when it proved impossible to develop an invulnerable basing mode for the MX did the government dismiss this fantasy. Although neither the original arguments for the significance of Minuteman vulnerability nor the more recent change of heart were entirely ingenuous, this episode illustrates how we can create problems for ourselves and present a picture that, if others accept it, can at least slightly diminish deterrence. On a more optimistic note, it also illustrates the possibilities for creativity that arise from the fact that the problems are not imposed on us by an external reality. The United States was facing a self-proclaimed deficiency that would have taken enormous sums to remedy, if it could have been remedied at all. Yet we were able to deal with it by issuing a report from a distinguished panel, a solution that not only cost much less money than the projected new systems, but also had fewer undesired international consequences.

Probably more dangerous than fears generated by supposed weaknesses in hardware are those produced by the elaboration of ever more complex scenarios. One can always generate finer and finer tactics of bargaining with nuclear weapons; one can always ask of any means of keeping the adversary in check, What if it fails? In the past, decision makers often dismissed certain dangerous situations as simply too implausible to be taken seriously. Thus, in 1924, Winston Churchill opposed the Admiralty's argument that more ships had to be built to meet the menace from Japan by arguing: "A war with Japan! But why should there be a war with Japan? I do not believe there is the slightest chance of it in our lifetime."[55] These judgments are difficult and, as this example shows, may be wrong. They are particularly hard to make today: the knowledge that world war would be a disaster and the limitations on our knowledge of how war might occur combine to defeat attempts to put bounds on our imagination. Brodie realized this: "All sorts of notions and propositions are churned out, and often presented for consideration with the prefatory words: 'It is conceivable that . . .' Such words establish their own truth, for the fact that someone has conceived of whatever proposition follows is enough to establish that it is conceivable. Whether it is worth a second thought, however, is another matter."[56] Of course, the penalties for an incorrect negative answer are enormous, but bounds must be established if we are not to pursue in ever more elaborate detail the implications of faulty premises.

WHY THERE HAS BEEN SO LITTLE CHANGE

The bulk of this essay has been devoted to arguing that there have been surprisingly few new ideas in nuclear strategy and that not all of the new ideas are good. Here, I note the possible sources of new ideas: changing technology, new knowledge, and our ability to learn from international events.

Technology is perhaps the most obvious possible source of new ideas. Indeed, one might think that it would not only encourage intellectual innovation, but also require it. Innovation in weapons systems is now the rule rather than the exception. But most of these changes, significant as they are in terms of military hardware and military spending, involve the substitution of one weapon for another, with little change in the projected missions. Furthermore, strategic thinking often anticipates the new weapons before they are produced.

Borden's book again is a marvelous example; it envisioned ICBMs, sea-launched missiles (he saw them on surface ships, not submarines), antimissile missiles, complex sensors, terminal guidance, and ABMs. Of course, the development of an ABM system that could protect cities would constitute a true revolution—or, more accurately, a counterrevolution—but there seems to be little chance of such a development in the foreseeable future.

A glance backward helps confirm the limited impact of technology. The development of accurate MIRVs, although disturbing to most strategists because they seemed to give incentives to strike first, have had much less influence on ideas about strategy or policy itself than most of us thought at the time. As Brodie pointed out, the strategic balance has too many components and is too stable to be greatly affected by even as substantial a change as this.[57] That a state could destroy more of the other's warheads than it would use up of its own does not provide a powerful enough incentive to strike first, in the face of the destruction that would likely result. Similarly, de-MIRVing, although desirable, would neither make the world much safer nor meet the requirements for deterrence set forth in current United States doctrine.

Some people have argued that new technology will lead to new concepts and policies by providing a new range of options.[58] The best example is the possibility of extremely accurate warheads, probably using some form of terminal guidance and perhaps carried by cruise missiles rather than ICBMs. (If accuracies were sufficiently high, conventional warheads could be used.) Such technologies would indeed allow the superpowers to do things they cannot do now, for example, destroy military or political control targets without doing great collateral damage. But would they drastically alter the current situation? I think not: new options will not alter mutual vulnerability. Although they would provide the possibility for ingenious limited wars, they could not do away with the overshadowing possibility of escalation. New technologies could make highly limited strikes possible but not safe. Would decision makers be likely to implement them in situations in which, under current circumstances, they would either not use nuclear weapons or else unleash a massive strike? Only if they thought the use of these options could terminate the war on acceptable terms. (Critics of limited options cannot have it both ways, however. If one argues that these options are close to useless, one cannot easily argue that statesmen would be quick to resort to them. It is possible, but a bit strained, to claim that although these options would not have the desired effects, statesmen would place confidence in them and so would behave incautiously.)

A second source of innovation is purely intellectual—new

ideas can be developed as analysts probe the problems more deeply and build on each other's work. No one would deny that this provides a partial description for what has happened. Progress is not automatic, however, and the generation of additional work does not automatically create new ideas.[59] Perhaps we lack the ingenuity to make basic intellectual breakthroughs; perhaps the ideas developed in the first fifteen years of the nuclear era have established the interesting lines of argument. In any event, more analysis does not automatically yield new thoughts.

The third possible source of new knowledge is history. Each year, we have a longer period of history from which to learn about nuclear strategies. Furthermore, each year we get new studies of earlier events, some based on new information and declassified documents. Much of this work is very significant and constitutes a major source of intellectual sustenance in our field. But we should recognize three inescapable limits to its value. First, there are only a few cases in which the use of nuclear weapons has been contemplated or threatened. Indeed, the Cuban missile crisis is probably the only instance in which war was seen as a real possibility. We also have no cases of such a confrontation when both sides had ample second-strike capability.[60] The lessons from these cases, furthermore, must remain especially ambiguous in the absence of evidence about the causes of Soviet behavior. Soviet documents would not clear up all of the mysteries—with every document declassified we still argue about the origins of World War I—but without them the room for disagreement is especially great. The ambiguity of the information feeds into the third limit on learning from the past. For scholars as well as decision makers, images of other countries and beliefs about politics are highly resistant to discrepant information. We all tend to assimilate new information into our preexisting beliefs. It is not surprising then that historical research has not led to agreement or to the development of many new ideas.

CONCLUSIONS

To argue for the revolutionary nature of nuclear weapons does not mean that all earlier, valid ideas no longer apply. The most important of these is Clausewitz's argument that war is not separate from politics, but rather is subordinate to it.[61] Military victory is not an end in itself; rather, it must serve political objectives, and we must remember that actions that bring military victory can also lead to political defeats. Thus, crossing the thirty-eighth parallel may have been the best way to destroy the North Korean army, but it was not an adequate means to reach American political objectives. Similarly, current strategic doctrine is constructed without attention to why the numerous options in the war plan could be expected to terminate the war; the crucial nexus between how the war would be conducted and how it might be stopped has been overlooked.

In the same way, the relationships between military means and political objectives rarely seem to figure in analyses of how World War III could start. Sometimes it is implied that the Russians are so strongly motivated to dominate the world that they would run high risks to reach this objective. This is the only way to make sense of the argument that it is vital for the United States to convince the Soviet leaders that their country would recover more slowly than the West in the aftermath of a war, as though the Russians might believe that this "advantage" would more than make up for the devastation of their homeland.[62] More often, however, little is said about what might drive the Russians to risk a war. In part, this silence can be explained by the knowledge that the detailed course of international politics cannot be predicted. It also reflects the difficulty in conjuring up a cause sufficient to start the conflict.

To focus on what political objective might warrant unprecedented Soviet aggression leads to the realization that desperation, rather than hope, is the likeliest cause of a nuclear war. Russia is more likely to be pushed into striking by the fear of what will happen if she does not, than pulled to war by the attraction of positive gains. The most potent fear, of course, is that war will start no matter what the Soviet Union does. This is not to deny that the expectation of gains can lead to more limited adventures, and that these in turn could lead to a major war. Even here, however, the final steps in the process probably would involve the belief that war will break out in the near future. Indeed, if war is seen as inevitable, deterrence cannot work—deterrence implies not only the threat of war if the other takes prohibited actions, but also the promise of peace if the other complies.

Security policy studies have paid careful attention to some parts of this problem, but not to others. As we noted, even though C^3 facilities are much more vulnerable than the weapons themselves, it seems extremely unlikely that even a well-designed attack on C^3 could prevent the victim from replying with devastating retaliation, although it could render impossible a strategy of controlled warfighting. Thus, only a state that was absolutely certain the other was about to strike should, if rationality rules, launch an attack. Fortunately, it is hard to imagine evidence that could produce such certainty.

Psychological dynamics compound the danger created by the weakness of C^3 systems, however. People try to avoid seeing things as entirely without hope; we all try to avoid seeing sharp trade-offs among our most central values.[63] So when people become desperate, they are likely to overestimate the chances of success of any way out of the situation.[64] A statesman who believes that war is inevitable will feel great psychological pressure to believe that a first strike could succeed. Furthermore, normal caution that acts as such a strong brake on wild behavior may not operate here. Even though most people are risk averse for gains, they seem to be risk acceptant for losses.[65] That is, although people will not run high risks to improve their situations, they may do so in the attempt to avoid losses. Thus, although it is certainly correct that under normal conditions Soviet—or American—leaders would not risk the fate of their country on one cosmic throw of the dice, they might do so in a deep crisis. If a decision maker believed that war was extremely likely and thought that striking the other's C^3 system would bring at least a small possibility of drastically limiting damage to his country, he might attack even if a risk-neutral calculus indicated that he should not.

Another aspect of crisis stability calling for attention is the need to make promises credible. The credibility of threats has been central to the field of strategic studies, but the state must also make credible its willingness to abstain from war and to respect the other's vital interests if the other cooperates. Although Schelling touched on this subject,[66] it has remained largely unexplored. Indeed, decision makers may not be aware of the need at all. As we saw, even as sophisticated a statesman as Dean Acheson did not think he had to convince the Chinese that a United Nations force on its Yalu border would not menace its security. Similarly, although statesmen in later crises realized that they had to guard against the danger that events would escape their control, they rarely appreciated the urgency of showing the other side that cooperation would be reciprocated. In part because they cannot empathize with the other side by realizing that it may fear that the state will strike first, decision-makers' reassurances tend to be perfunctory at best. Scholars have not done much better; we do not have sustained analyses of how promises can be made credible, what devices or tactics might be employed, the barriers to the accurate perception of such commitments, and the conditions under which leaders have thought creatively about the problem.

The final neglected aspect of how war might start is related to the previous point: neither statesmen nor scholars have paid sufficient attention to the conditions that produce the beliefs that the costs of remaining at peace are unbearable or that war is inevitable. Eroding the other's power too much may produce short-run gains, but it may also push the other closer to war. For, if either side believes that remaining at peace cannot preserve its core values, it may take high risks to reverse the course of events.

As long as we live in an era of mutual vulnerability, making the other side less secure is not likely to redound to the state's long-run benefit. As Michael May argues: "It would lessen, not improve, Soviet security if the Soviet Union were to 'win' in the Caribbean, particularly if it won big and began to be a military factor in, say, Mexico. The extraordinary insecurity created in the United States would greatly overbalance whatever conventional military advantages might accrue."[67] Similarly, as much as the United States would like to see democratic regimes established in East Europe, would such an eventuality increase or decrease American security? At times, American statesmen have acknowledged this problem, but as an afterthought rather than a central focus.

What is most important, then, may be the relationships between the two superpowers' security. On the one hand, to the extent that both sides adopt military doctrines that stress the ability to deny the other any military advantage at any level of conflict, the security dilemma is unavoidable. That is, the two sides' security requirements will directly clash—the only way one side can feel secure is to make the other less secure. On the other hand, the realization that making the other side desperate is very dangerous implies that neither side can be secure if the other is very insecure.[68] It indicates, then, that in the nuclear era the only security may be mutual security.

NOTES

1. Bernard Brodie, *The Absolute Weapon* (New York: Harcourt Brace, 1946). Of course, this is not to argue that Brodie's ideas were consistent and unchanging. See Barry Steiner, "Using the Absolute Weapon: Early Ideas of Bernard Brodie on Atomic Strategy," *Journal of Strategic Studies* 7 (December 1984), pp. 365–93. I am grateful to McGeorge Bundy for having pointed out to me that, in November 1945, *Time* listed twelve points that, like Brodie's arguments, have proved to be remarkably prescient. Among them were the following: "Atomic weapons will overshadow peacetime uses of atomic energy"; "No military or scientific defense can be expected"; "Much larger atomic charges are in prospect"; "No big secret protects the atomic bomb"; and, most important, "Out-producing the enemy is not much advantage in atomic warfare. Two hundred may be better than 100 bombs, but 10,000 is no better than 5,000, because 5,000 would destroy all important targets in a country." (*Time*, November 12, 1945, p. 28.)

2. For further discussion of the nuclear revolution, see Michael Mandelbaum, *The Nuclear Revolution* (New York: Cambridge University Press, 1981); and Robert Jervis, *The Illogic of American Nuclear Strategy* (Ithaca, N.Y.: Cornell University Press, 1984), chapter 1.

3. Glenn Snyder, *Deterrence and Defense* (Princeton, N.J.: Princeton University Press, 1961).

4. David Rosenberg, "American Atomic Strategy and the Hydrogen Bomb Decision," *Journal of American History* 66 (June 1979), pp. 71–7.

5. William Borden, *There Will Be No Time* (New York: Macmillan, 1946). I am grateful to James King, *The New Strategists* (unpublished manuscript) for stressing the significance of this book. It should be noted that Borden argued that nuclear weapons did bring about a revolution, but he is able to assert this only by misinterpreting crucial aspects of World War II.

6. "Is SALT II a Fair Deal for the United States?" (Washington, D.C.: Committee on the Present Danger, 1979), p. 6.

7. Although this essay will concentrate on ideas about strategic policy rather than on the policy itself, some parallels between the two should be noted. John Gaddis has argued that American policy can be divided into periods characterized by symmetry or asymmetry. By the former, he means policies that seek to meet the Soviet threat at each level of violence; by the latter, those that are built on the premise that nuclear weapons are sufficient to protect a wide range of American values. In other words, symmetric policies seek deterrence by denial, at least for levels of violence below all-out war, whereas asymmetric policies fully accept the nuclear revolution and utilize deterrence by punishment. Gaddis's explanation has little to do with the power of abstract ideas; rather, he sees the cause in the differences in administrations' willingness to support high defense budgets. See *Strategies of Containment* (New York: Oxford University Press, 1962).

8. For such a history, see Lawrence Freedman, *The Evolution of Nuclear Strategy* (New York: St. Martin's, 1981). As Freedman notes, "Much of what is offered today as a profound and new insight was said yesterday; and usually in a more precise and literate manner" (p. xv).

9. Caspar Weinberger, *Annual Report to the Congress, F.Y. 1987* (Washington, D.C.: Government Printing Office, 1986), p. 73.

10. Department of Defense, *Annual Report, F.Y. 1981* (Washington, D.C.: Government Printing Office, 1980), p. 63; ibid., p. 65; ibid., p. 67; and Harold Brown, remarks prepared for delivery before the Council on Foreign Relations and the Foreign Policy Association, April 5, 1979; Department of Defense, *Annual Report, F.Y. 1980* (Washington, D.C.: Government Printing Office, 1979), p. 77.

11. Thomas Schelling, *Arms and Influence* (New Haven, Conn.: Yale University Press, 1966), pp. 190–2.

12. Weinberger, *Annual Report, F.Y. 1987*, p. 13.

13. Fred Ikle, quoted in Richard Berstein, "Americans, at Forum in France, Put forward a Case for 'Star Wars,' " *New York Times*, October 21, 1985, p. B10. Also see Samuel Huntington, "The Renewal of Strategy," in Huntington, ed., *The Strategic Imperative: New Policies for American Security* (Cambridge: Ballinger, 1982), p. 49.

14. Schelling, *Arms and Influence*, p. 98.

15. Bernard Brodie, *War and Politics* (New York: Macmillan, 1973), p. 412.

16. Bernard Brodie, *Strategy in the Missile Age* (Princeton, N.J.: Princeton University Press, 1959).

17. "Transcript of Press Interview with President at White House," *New York Times*, March 30, 1983.

18. Jonathan Schell, *The Fate of the Earth* (New York: Knopf, 1982); Schell, *The Abolition* (New York: Knopf, 1984).

19. But see the argument between President Eisenhower and Secretary of State Dulles in the National Security Meeting of June 23, 1954, as recorded in Department of State, *Foreign Relations of the United States, 1952–1954*, vol. II, *National Security Affairs*, Part 2 (Washington, D.C.: Government Printing Office, 1984), pp. 1467–70.

20. Hans Morgenthau, "The Fallacy of Thinking Conventionally about Nuclear Weapons," in *Arms Control and Technological Innovation*, ed. David Carlton and Carlo Schaerf (New York: Wiley, 1976), pp. 256–64.

21. Robert McNamara, "The Military Role of Nuclear Weapons," *Foreign Affairs* 62 (Fall 1983), p. 68.

22. Jervis, *The Illogic of American Nuclear Strategy*.

23. Herbert Butterfield, *History and Human Relations* (London: Collins, 1951); John Herz, "Idealist Internationalism and the Security Dilemma," *World Politics* 2 (January 1950), pp. 157–80.

24. For a discussion of the conditions under which deterrence theory or the spiral model is more appropriate, see Robert Jervis, *Perception and Misperception in International Politics* (Princeton, N.J.: Princeton University Press, 1976), chapter 3.

25. For a discussion of this issue, see Richard Ned Lebow, "The Soviet Offensive in Europe: the Schlieffen Plan Revisited?" *International Security* 9, (Spring 1985), pp. 44–78. Of course, fortifications are not entirely defensive because they allow a state to protect its border with a relatively small number of soldiers, thus leaving a large part of its army free to take the offensive. Similarly, states may seek to take the offensive, not because they are aggressive and want to acquire more territory, but because they seek to deter others; see Samuel Huntington, "Conventional Deterrence and Conventional Retaliation in Europe," *International Security* 8 (Winter 1983–84), pp. 32–56.

26. For a nice restatement of this position, see Thomas Schelling, "What Went Wrong with Arms Control?" *Foreign Affairs* 64 (Winter 1985–86), pp. 219–33.

27. See Robert Jervis, "Collaboration under the Security Dilemma," *World Politics* 30 (January 1978), pp. 167–214; Robert Axelrod, *The Evolution of Cooperation* (New York: Basic Books, 1984); Robert Keohane, *After Hegemony* (Princeton, N.J.: Princeton University Press, 1984); and the special issue of *World Politics* 38, ed. Kenneth Oye (October 1985). For criticisms of this perspective, see Harrison Wagner, "The Theory of Games and the Problem of International Cooperation," *American Political Science Review* 77 (June 1983), pp. 330–46; also see Hayward Aiker, Jr., and Richard Ashley, *Anarchy, Power, Community* (forthcoming).

28. See the special issue of *International Organization* 36, ed. Stephen Krasner (Spring 1982).

29. Even though there are important similarities between the two realms, the nature of international security politics means that cooperation is more difficult to achieve in that area than in political economy. See Jervis, "Security Regimes," *International Organization* 36 (Spring 1982), pp. 358–60; and Charles Lipson, "International Cooperation in Economic and Security Affairs," *World Politics* 37 (October 1984), pp. 1–23.

30. Thomas Schelling, *The Strategy of Conflict* (New York: Oxford University Press, 1963), pp. 134–5.

31. I have discussed other examples of this pattern in "From Balance to Concert: A Study of International Security Cooperation," *World Politics* 38 (October 1985), pp. 76–8. Also see P. J. V. Rolo, *Entente Cordiale* (New York: St. Martin's 1969), p. 49.

32. Leo Sartori, "Will SALT II Survive?" *International Security* 10 (Winter 1985–86), pp. 164–5.

33. *New York Times*, May 23, 1984.

34. Weinberger, *Annual Report, F.Y. 1987*, p. 76.

35. F. H. Hinsley et al., *British Intelligence in the Second World War*, vol. II (New York: Cambridge University Press, 1981), p. 117.

36. Axelrod, *The Evolution of Cooperation*. Also see Anatol Rappoport, "Editorial Comment," *Journal of Conflict Resolution* 29 (December 1985), p. 621.

37. George Downs, David Rocke, and Randolph Siverson, "Arms Races and Cooperation," *World Politics* 38 (October 1985), pp. 134–43.

38. Jervis, *Perception and Misperception in International Politics*, pp. 68–75, 349–55.

39. Alexander George, *Managing U.S.–Soviet Rivalry* (Boulder, Colo.: Westview, 1983); Raymond Garthoff, *Détente and Confrontation* (Washington, D.C.: Brookings, 1985); Roger Bullen, *Palmerston, Guizot and the Collapse of the Entente Cordiale* (London: Athlone, 1974); Gordon Craig, "The System of Alliances and the Balance of Power," in *New Cambridge Modern History*, vol. X, *The Zenith of European Power, 1830–70*, ed. J. T. P. Bury (New York: Cambridge University Press, 1960), pp. 254–7.

40. Quoted in John Spanier, *The Truman–MacArthur Controversy and the Korean War* (New York: Norton, 1965), p. 97.

41. Robert Axelrod, *The Evolution of Cooperation*, pp. 184–5.

42. George, "Political Crises," in *The Making of America's Soviet Policy*, ed. Joseph Nye (New Haven, Conn.: Yale University Press, 1984), p. 150.

43. Jervis, *Perception and Misperception in International Politics*, pp. 62–84.

44. For a further discussion, see Michael Nicholson, *Oligopoly and Conflict: A Dynamic Approach* (Toronto: University of Toronto Press, 1972), chap. 6; Michael Taylor, *Anarchy and Cooperation* (London: Wiley, 1976), p. 96; and Joanne Gowa, "Anarchy, Egoism, and Third Images," *International Organization* 40 (Winter 1986), pp. 182–3.

45. See, for example, Amatai Etzioni, *The Hard Way to Peace* (New York: Collier, 1962); Jerome Frank, *Sanity and Survival in the Nuclear Age* (New York: Random House, 1967); Charles Osgood, *An Alternative to War or Surrender* (Urbana, Ill.: University of Illinois Press, 1962); Ralph White, *Nobody Wanted War* (Garden City, N.Y.: Doubleday, 1968). White's more recent book,

Fearful Warriors (New York: Free Press, 1984), combines the strengths of the earlier work with a good appreciation of the nature of international politics and the danger of aggression and failures of deterrence.

46. Paul Schroeder, *Austria, Great Britain, and the Crimean War* (Ithaca, N.Y.: Cornell University Press, 1972); Richard Ned Lebow, *Between Peace and War* (Baltimore: Johns Hopkins University Press, 1981); Robert Jervis, Richard Ned Lebow, and Janice Stein, *Psychology and Deterrence* (Baltimore: Johns Hopkins University Press, 1985).

47. Bernard Brodie, "On Clausewitz: A Passion for War," *World Politics* 25 (January 1973), pp. 288–308.

48. See the essays in *Strategic Nuclear Targeting*, ed. Desmond Ball and Jeffery Richelson (Ithaca, N.Y.: Cornell University Press, 1986).

49. John Steinbruner, "An Assessment of Nuclear Crises," in *The Danger of Nuclear War*, ed. Franklyn Griffiths and John Planyi (Toronto: University of Toronto Press, 1979), pp. 34–49.

50. Charles Perrow, *Normal Accidents* (New York: Basic Books, 1984).

51. Two case studies of military alerts are presented in Scott Sagan, "Nuclear Alerts and Crisis Management," *International Security* 9 (Spring 1985), pp. 99–139. For discussion of similar problems in the prenuclear era, see Jervis, *Perception and Misperception in International Politics*, pp. 329–42.

52. Jack Snyder, *The Ideology of the Offensive* (Ithaca, N.Y.: Cornell University Press, 1984).

53. Jervis, *The Illogic of American Nuclear Strategy*, pp. 126–9; Jervis, "Psychological Aspects of Crisis & Stability" (unpublished manuscript); Richard Ned Lebow, *Nuclear Crisis Management: A Dangerous Illusion* (Ithaca, N.Y.: Cornell University Press, forthcoming).

54. David Rosenberg, "The Origins of Overkill: Nuclear Weapons and American Strategy, 1945–1960," *International Security* 7 (Spring 1983), pp. 3–71; Desmond Ball, "U.S. Strategic Forces: How Should They Be Used?" *International Security* 7 (Winter 1982–83), pp. 31–60.

55. Quoted in Martin Gilbert, *Winston S. Churchill, 1922–1939*, vol. 5 (London: Heinemann, 1976), p. 76.

56. Bernard Brodie, "The Development of Nuclear Strategy," *International Security* 2 (Spring 1978), p. 83.

57. Bernard Brodie, "On the Objectives of Arms Control," *International Security* 1 (September 1976), pp. 17–36.

58. Henry Rowen and Albert Wohlestetter, "Varying Response with Circumstance," in *Beyond Nuclear Deterrence*, ed. Johan Holst and Uwe Nerlich (New York: Crane, Russak, 1977), pp. 225–38.

59. For a somewhat parallel discussion of progress—or lack of progress—in the quantitative analysis of international politics, see James Rosenau, ed., *In Search of Global Patterns* (New York: Free Press, 1976).

60. The United States may have acted as though the Russians had such a capability in 1962. See Fred Kaplan, *The Wizards of Armageddon* (New York: Simon and Schuster, 1983), pp. 298–306.

61. Brodie stressed this point and argued, I think correctly, that although it is obvious, it is often overlooked. See *War and Politics*, chap. 1.

62. For a further discussion, see Jervis, *The Illogic of American Nuclear Strategy*, pp. 59–63.

63. Jervis, *Perception and Misperception in International Politics*, pp. 128–42.

64. The way these dynamics can lead to war has been demonstrated in Lebow, *Between Peace and War*, and Jervis, Lebow, and Stein, *Psychology and Deterrence*.

65. For a summary of the evidence and arguments on this point, see Daniel Kahneman and Amos Tversky, "Choices, Values, and Frames," *American Psychologist* 39 (April 1984), pp. 341–50.

66. Schelling, *Strategy of Conflict*, pp. 43–4, 120, 129, 131–4.

67. Michael May, "The U.S.–Soviet Approach to Nuclear Weapons," *International Security* 9 (Spring 1985), p. 150.

68. This question was discussed, although not quite in these terms, by the Eisenhower Administration: U.S. Department of State, *Foreign Relations of the United States, 1952–1954*, vol. II, part 1 (Washington, D.C.: Government Printing Office, 1984), pp. 529–30, 757.

CHAPTER 3

Cultural Barriers to an Acceptance of Deterrence

GEORGE H. QUESTER

As we enter the fifth decade after Hiroshima, the concept of deterrence is under a somewhat unfair attack from both left and right. Critics from the left assail it as justifying too many weapons systems, or as assuming too much in the way of human rationality for heading off an ultimate thermonuclear cataclysm.[1] Critics from the right claim that the Soviet Union does not share Western concepts or views here and perhaps is not as frightened of nuclear war as we are in the West; deterrence reasoning is thus allegedly bound to lead to disappointments on detente and arms control and is bound to leave such valuable places as Western Europe exposed, now that the Soviet nuclear arsenal has grown to match or surpass that of the United States.[2]

The central ideas of deterrence are assailed as being somehow disproven and bankrupt, or instead as irrelevant. The vendors of such ideas seem to have little or nothing new to offer, even though no real solutions have yet been achieved through application of such deterrence reasoning. What is taught as a confrontation of nuclear deterrence, in university classes across the United States, is now shrugged off as the abstraction of academics, allegedly unrelated to real policy in either Washington or Moscow.

Yet, the latter criticism may be a little closer to the mark, in a way that negates the former. The abstract logic of nuclear deterrence, in its extreme form that we should welcome the vulnerability of our own civilians to an adversary's retaliatory nuclear attack (already morally abrasive in that it suggests that *we* should be targeting *their* civilians), has all along had a very uphill battle to win acceptance, for reasons that may say little or nothing about the validity of the theory. If such reasoning has not been fully accepted in Washington, or in Moscow, or in Omaha, can we fault its logic for failing to solve our problems, and for failing to head off our disappointments?

This essay outlines some moral, philosophical, and psychological factors, which have all along barred any complete acceptance of the logic of massive retaliation and nuclear deterrence, and then describes the ways in which this incompleteness of the intellectual revolution has caused some serious problems.

In the immediate aftermath of the introduction of nuclear weapons at Hiroshima and Nagasaki, as most other Americans were celebrating the good news of the Japanese surrender that terminated World War II so surprisingly early, two scholars working at Yale University independently began writing what could be seen as the poles of our debate for the next four decades on deterrence and its alternatives. Bernard Brodie stated the case for deterrence, the case that the retaliatory countervalue destruction of civilian targets would now hold back the use of such weapons by each side and perhaps other forms of warfare as well, in the book he edited, *The Absolute Weapon*.[3] William Borden presented the more traditional form of weapons evaluation in a very different book, *There Will Be No Time*, stressing the urgency of military preemption, of destroying the other side's military capability for fighting a war, before its military forces destroyed ours.[4]

A great deal of the dissatisfaction expressed today regarding deterrence relates to this second, more traditional, war-fighting view, arguing that such an emphasis on disarming the enemy would be more moral, or that it would be more conducive to American national interests (or, at the very least, that it is the strategic perspective that continues to guide the Soviets, regardless of our own views). Each of these contemporary arguments, in effect, presupposes that mutual deterrence has indeed been adopted as the American strategic perspective, that Brodie's perspective was adopted ahead of Borden's; but has this acceptance of mutual assured destruction (MAD) actually been so total or so clear? It may be less appropriate to criticize the effects of a strategy that has never been fully accepted.

SOME BIASES AGAINST DETERRENCE

One can think of at least four reasons that Borden's more traditional interpretation of the significance of nuclear weapons was likely to persist, the interpretation that treats these weapons like any other weapon—intended to be used, intended to blunt the enemy's use of the same kinds of weapons.

First, the reluctance to accept mutual deterrence could have reflected simple inertia. Old ideas die hard. Old soldiers will not reject the wisdom of Clausewitz or Mahan overnight, simply because of some nuclear explosions or some writings of professors. The writers analyzing future aerial combats after World War I often outlined attacks on cities (many cities on the same day) that look a great deal like today's prospect of nuclear warfare, patterns of attack much more horrendous than those actually experienced in World War II. Yet writers such as Giulio Douhet and J. F. C. Fuller did not then complete the logical pattern to predict any mutual deterrence, but rather merely foresaw opposing bomber forces racing past each other to devastate each other's cities.[5]

Second, such a counterforce use of nuclear weapons seems much less immoral, less contrary to all of Western philosophical tradition, less inclined to produce indictments for war crimes or pangs of conscience for those who take seriously the statements of their bishops. As ordinary newspaper editorials demonstrate, military officers are not alone in being reluctant to accept the inverted (perverted?) moral logic of nuclear deterrence.

Third, it has seemed possible all along that our counterforce attacks might one day be effective in a *splendid first strike,* shielding North America against all destruction as we won a war. And we have had to be on our guard lest the Soviets uncover some opportunities for a splendid first strike from their side, stripping us one morning of all of our ability to retaliate against them.

Fourth, turning to more venal arguments, it is also very plausible that air forces might hope to increase their budgets and manning by stressing the counterforce role ahead of countervalue deterrence. The former requires more bombers and missiles than the latter. Stressing the counterforce threats of an opposing side's bombers and missiles, moreover, compounds the enhancement of one's own budgetary arguments. "There will be no time" to match their procurements later, so we must get our procurements now. Just as navies have built up their constituencies in the past (and present) by emphasizing the threats of opposing navies, as well as the capabilities of their own proposed fleets, an emphasis on bomber gaps and missile gaps would increase the numbers of bombers and missiles we would be inclined to purchase.[6]

The last of such arguments are thus ad hominem, accusing the advocates of strategies other than deterrence of a conscious or unconscious pursuit of career advancement, or of simple inertia, in an inability to comprehend how nuclear weaponry has changed strategy. Yet others of these arguments are more worthy, suggesting that directing nuclear threats against civilian populations would produce too many moral side effects, or perhaps that deterrence theories are too premature in assuming what would be called *assured destruction*. What if the nuclear destruction of cities would not be enough to force a determined regime to surrender, as long as its forces were winning victories otherwise on the battlefields? The public's reaction to the

1946 Bikini test of atomic bombs was surprise at how little destruction was inflicted, rather than at how much; the event was memorialized in the naming of a swimsuit.[7]

The debate sometimes intertwines issues in ways that make it very difficult to sort out the sincerity or the bureaucratic self-service of the cases. In a standard kind of infighting, which characterizes competition among the different components of the United States armed forces, the issues of morality of targeting have surfaced several times. At the end of the 1940s, the United States Navy, trying to win greater access to the limited stockpile of nuclear weapons and to appropriations for the delivery systems to carry them, accused the Air Force, in its emphasis on strategic bombing, of planning for an immoral countervalue, countercivilian war. Because Navy aircraft carriers would be directing their missions more at tactical targets closer to the frontlines, they might be hitting fewer civilians in their missions, thus winning wars by actually defeating the enemy's military forces, rather than by destroying his willingness to fight by flattening his cities.

To show that such moral standards can be rather transitory, particularly for a military service intent on advancing its own share of the budget, the same Navy by the end of the 1950s was endorsing a very explicit philosophy of *finite deterrence*, aimed at deterring wars by an unignorable threat to Soviet cities. The Navy was now pushing its Polaris program for submarine-based ballistic missiles carrying H-bomb warheads, a superbly stabilizing retaliatory force, because any enemy intent on a disarming splendid first strike would have so much difficulty detecting the location of these submarines. The Polaris missiles fired from such submarines would have suffered from great inaccuracies, making them relatively ineffective for any traditional role involving a precise attack on explicitly military targets, but they were more than adequate for second-strike assured destruction. Admiral Raborn, the project manager for the Polaris program, was reportedly asked whether the poor accuracy of the submarine-based missile would be a major problem. "If it even hits dry land, it's accurate enough," was his answer. The Navy was now defending its program citing the wars it would prevent, because classic civilian targets would suffer so badly.

ACCEPTANCES OF DETERRENCE

Some or all of the influences just outlined may account for the continuing treatment of nuclear weapons as military weapons, rather than as instruments of retaliation that, once they have been used, merely take all the point out of winning a war. Yet the alternative view, the "new" view of mutual assured destruction, has also won endorsements again and again since 1945, with even some forerunners extrapolating from a different kind of warhead in the years before 1939. What we have seen for the years since Nagasaki is thus a form of continuing debate, a debate that may finally be brought to an end with the formulation of *nuclear winter* (but which may not), a debate that perhaps should have been ended some two decades ago.

Bernard Brodie's book of 1946 was remarkably free of any notions of applying the American monopoly at the time, with a view to preserving the monopoly, by waging a preventive war against the Soviet Union or dictating terms of disarmament to Stalin. Rather, it presciently looked forward to the day when Russia would have weapons as well. Brodie assumed that a comparison of numbers in the nuclear stockpiles on the two sides would not make a great deal of difference, for a stockpile on the Soviet side sufficient to direct some atomic warheads against New York and a few more American cities would be enough to deter our own use of the United States nuclear stockpile.

A similar logic is evident in the internal NSC–68 document, commissioned by President Truman to examine the consequences of the Soviet atomic test detected in 1949, a document put together under the general direction of Paul Nitze.[8] Still assigning a great deal of importance to counterforce exchanges, the study, reflecting many of the confusions and debates we are outlining here, at the same time stipulated that Soviet possession of even a rudimentary nuclear stockpile might deter any use of the American nuclear arsenal, so that substantial reinforcements of conventional forces might be required to check Soviet expansion.

Henry Kissinger's 1957 book *Nuclear Weapons and Foreign Policy,* outlined a very similar argument in chapter 5, attaching an "absolute weapon" status to the damage Soviet nuclear weapons could do to American cities. In chapter 6 of the same book, Kissinger then startled his contemporaries by proposing tactical nuclear weapons, rather than a reliance on American conventional weapons, as the counter to a Soviet conventional force, limited war attack, working from the premise that it would not be so difficult to control and limit the use of nuclear weapons in a limited war.[9]

Kissinger joined the mainstream of limited war theorists in his next book, *The Necessity for Choice,*[10] published in 1961, conceding that it would be quite difficult to control and limit the use of nuclear weapons once they had been introduced into any battlefield use at all, endorsing instead the proposals for buildups in conventional forces. For our purposes, the important point here is that Kissinger, in these first two books (and indeed in his next book and in his statements made as Secretary of State), endorsed the interpretation of strategic nuclear weapons by which their countervalue impact was paramount, so paramount or absolute that any workings out of a counterforce war-fighting exchange would be swamped by the fact that all the major cities of the United States would have been destroyed, no matter who had "won" the war.

Kissinger's third book, *The Troubled Partnership* (1965), seemed on first glance to revert to the policies endorsed in his first, as tactical nuclear weapons were to be kept deployed in West Germany to prevent a Soviet conventional force aggression against NATO.[11] But the logic was very different. Rather than claiming that such tactical nuclear weapons could be used without escalation, Kissinger now argued that escalation was indeed likely; precisely this prospect would continue what had deterred and dissuaded Moscow from aggressions ever since World War II.

In all three of these books, Kissinger was wrestling with the knottiest problem of post-1945 American military strategy—not how to prevent the Soviet Union from attacking the United States, but how to deter the Soviet Union from attacking Western Europe. Once the Soviets had any nuclear weapons at all, according to Kissinger (just as according to Brodie), the overarching comparison of strategic nuclear forces was in danger of simply cancelling itself out, in a pattern of mutual assured destruction, mutual deterrence. Therefore, if it would be madness for the United States to escalate to all-out nuclear war when only Hamburg or Frankfurt was attacked, how did one dissuade Moscow from sending tanks into Frankfurt or Hamburg? Kissinger offered three solutions in three books; but none of these solutions questioned the premise that a nuclear threat against every American city of more than five hundred thousand population was sufficient to be absolute.

When discussing the terms of the SALT I agreement in 1974, Kissinger again made the absolute weapon point very eloquently when scoffing at United States Air Force arguments about the significance of nuclear strategic superiority. "What in the name of God is strategic superiority? What is the significance of it, politically, militarily, operationally, at these levels of numbers? What do you do with it?"[12]

Shifting with the politics of the time, or being swayed by the continued vitality of arguments focusing on the more military counterforce aspects of nuclear weaponry, Kissinger had reversed himself on this most basic of points by September 1982, when he delivered a controversial speech at Bruxelles arguing that something had indeed been lost when Soviet missile totals had climbed by one measure or another to match or surpass those of the United States.[13]

In addition to flatly contradicting the 1974 quote noted above, this sense of loss flatly contradicts the pessimistic premises of his 1956 book. All of what nuclear superiority could amount to had already been lost, by the 1956 analysis, matching Brodie's 1946 analysis and the premises of NSC–68, once a single bomb in Moscow's arsenal could destroy New York, another Washington, and another Los Angeles.

To sharpen the point here, one view (that of mutual assured destruction) would argue that everything significant was lost once the nuclear monopoly was broken, with the lingering American superiority in 1950 unexploitable for anything, and with a new Soviet superiority in the 1980s also unexploitable. Two important summations and statements of this kind of position were offered at the end of the 1960s by Wolfgang Panofsky and McGeorge Bundy.[14] The viewpoint which instead sees nuclear weapons mostly as weapons, rather than as countercivilian instruments of retaliation, will be rather more concerned about other transitions, from United States superiority to parity or essential equivalence, and then to Soviet superiority.

COUNTERFORCE PLANS, COUNTERVALUE RESULTS

Historians examining the declassified nuclear war plans of the late 1940s, 1950s, and 1960s often declare themselves amazed at how much these plans were phrased in counterforce rather than countervalue terms, amid extensive lists of targets and extensive commitments of nuclear warheads.[15] After decades of hearing civilian strategic analysts and professors describing a pattern of mutual deterrence and mutual assured destruction, based on countervalue targeting of urban areas, these war plans are cited as evidence of some deep misunderstanding on our campuses and some serious anomalies in the government war planning. Because a great many civilians would still have been killed in such counterforce bombing campaigns as planned in Omaha, moreover, the war plans are also portrayed as showing great folly and hypocrisy in the selection of targets by headquarters.

In light of what we have outlined here, such gasps of astonishment from the historians are not so well taken. It was perfectly natural that General Lemay and his targeting staffs for the Strategic Air Command would still be locked into counterforce thinking on the use of nuclear weapons. Those who had doubts about whether such bombings were absolute, whether bombing would surely be sufficient to deter war, could feel that some attention needed to be paid to winning such a war if it occurred. Germany did not surrender when Hamburg and Dresden were incinerated, but only when United States ground forces had advanced to meet Soviet forces coming from the east. To repeat the list noted earlier, such counterforce thinking also justified greater expenditures, and it was more in keeping with Western moral thinking and American traditions in air war planning. All the United States bombing campaigns of World War II, over Germany and Japan, had officially been declared efforts to destroy the enemy's *ability* to wage war, not to punish German or Japanese civilians; even the fire bombing of Tokyo and the atomic bombings of Hiroshima and Nagasaki were somehow squeezed into this category.

For many analysts, the important point is thus not what moral frame the Omaha target planners chose to adopt, but the inevitable fact that a great deal of countervalue damage would be inflicted in such American counterforce strikes, so much that Moscow could hardly ignore the assured destruction that would ensue. Just as in World War II and other bombing campaigns, the civilian suffering that the Russians would experience would be collateral damage, but one always found a number of military targets inside each Soviet city. If this can be indicted as some kind of hypocrisy, shielding American military officers and civilian leaders from thinking of themselves as war criminals, it was a very useful hypocrisy, making certain that a nominal counterforce philosophy produced reliable countervalue results.

As long as the counterforce war-fighting plans coming out of Omaha could not hope to make New York or Washington or San Francisco secure against a Soviet nuclear second-strike attack (a responsive Soviet attack that also might come in counterforce military target clothes, but have a decidely countervalue impact), nothing about these apparent anomalies or hypocrisies or contradictions could upset the conclusion that a mutual deterrence was in place, based on mutual capabilities for absolutely unignorable destruction.

IF DETERRENCE IS DELICATE, IS IT ILLEGITIMATE?

One could, of course, have conceded that A-bombs or H-bombs were absolute weapons, unignorable if poised to hit one's cities in a last-gasp retaliation, and still have concluded that counterforce strikes and war fighting had some meaning, *if* there was any prospect of such counterforce strikes catching almost all of the other side's retaliatory warheads before they could be fired. Rather than simply scoffing at the Bikini test, or simply nostalgically noting that wars had continued to be an exchange of military operations in the actual conduct of World War II (as well as in the scenarios of Douhet), this would assume that the bomber would not always get through, if the other side's bombers carried off their counterforce strikes early and effectively enough.

Albert Wohlstetter published an article in 1959, "The Delicate Balance of Terror," extrapolating from an earlier classified RAND Corporation study he had directed for the United States Air Force on possible vulnerabilities in the basing of its strategic bombers; the article captured a great deal of attention simply because of the message so aptly summarized in its title.[16] In part because the Kremlin leadership might be callous about the suffering inflicted in a *few* nuclear or thermonuclear strikes at Soviet cities, and in larger part because there were allegedly many other ways in which a clever Soviet surprise attack might thin out and preempt any American strategic nuclear retaliation, Wohlstetter argued: that mutual deterrence was not so certain; that (in the next decade's phrasing) mutual assured destruction was hardly so assured; and that the balance of terror was delicate rather than certain. Wohlstetter directed his argument against what he saw as a widely held conclusion about the destructiveness of a nuclear exchange and the consequent inutility of nuclear weapons. He was assailing an assumption that deterrence would take care of itself, the premise one might impute to Brodie and then to Kissinger, the premise that seemed all the more self-evidently plausible as the United States reacted to the Eniwetok H-bomb test (no one named a swimsuit for Eniwetok).

In the face of incipient, feared bomber gaps, future missile gaps, and the possible vulnerabilities to first strike in bomber versus bomber or missile versus missile confrontations, Wohlstetter's message in part (but only in part) seemed like a sophisticated endorsement of the gut reactions military professionals had carried forward from the years of World War II and before; that aiming at the enemy's military force still could make a great deal of sense (or, at the least, worrying about the enemy aiming at your military forces made a great deal of sense).

Besides his scenarios of how the United States might be vulnerable to a Soviet counterforce first strike in the 1950s or 1960s (scenarios that amounted to poor predictions in the later 1960s and 1970s, when seemingly secure solid-fuel missiles were deployed in siloes and on board submarines on both sides, but that have captured new life again in the 1980s with worries about a *window of vulnerability*), Wohlstetter included a number of other interesting observations in his article. He deprecated the value of forward-based nuclear forces as a solution to the American strategic vulnerability problem, dismissing any alleged Soviet difficulties in the timing of strikes at different targets at different ranges, as well as the alleged military advantages of such forward-based nuclear systems for fighting a nuclear defense of Western Europe. In an argument to be expanded substantially in a later article entitled "Nuclear Sharing: NATO and the N + 1 Country,"[17] Wohlstetter also deprecated any alleged advantages of nuclear proliferation (France being the country in mind at the time).

Illustrating a basic acceptance of the logic of mutual deterrence (despite all the worries he was advocating on its attainability), Wohlstetter also offered a short discussion of yet another disadvantage of forward deployment—it might be seen in Moscow as threatening a United States counterforce first strike.

> There is one final consideration. Missiles placed near the enemy, even if they could not retaliate, would have a potent capability for striking first by surprise. And it might not be easy for the enemy to discern their purpose. The existence of such a force might be a considerable provocation and in fact a dangerous one in the sense that it would place a great burden on our deterrent force which more than ever would have to guarantee extreme risks to the attacker—worse than the risks of waiting in the face of this danger. When not coupled with the ability to strike in retaliation, such a capability might suggest—erroneously, to be sure, in the case of the democracies—an intention to strike first. If so, it would tend to provoke rather than to deter general war.[18]

It would be interesting to see whether Wohlstetter would want these phrases to be cited today in an argument against the deployment to West Germany of the Pershing II missile, allegedly so worrisome because it can reach Moscow in only about eight minutes traveling time.

Echoing arguments explicated much more clearly in a work of Oscar Morgenstern, *The Question of National Defense*,[19] Wohlstetter thus endorsed an avoidance of the most meaningful of counterforce weaponry for the United States, weaponry that could seem to be about to strip the adversary of his retaliatory hold on American cities. The Wohlstetter argument was decidely *not* any simple skepticism about countervalue impact, freeing the military professional to return undistracted to planning for a counterforce victory. Rather, it was a message that those who wished to focus undistractedly on countervalue retaliation could not have the luxury of doing so, for the enemy might be able to sneak into a meaningful counterforce capability; at the least, we would continually have to calculate the confrontation of such possible capabilities, lest we wake up some morning the victims of a Soviet splendid first strike.

Wohlstetter's argument was complicated and profound, far more than a mindless discussion of "wars fought as in the good old days." Such simple nostalgia might have accounted for some earlier (Douhet, J. F. C. Fuller, Borden) projections of horrendous air power that somehow never got to the possibility (the necessity?) of mutual deterrence. Wohlstetter endorsed the need for deterrence but raised doubts about whether nuclear weapons were quite terrible enough to deter everyone and about the survivability of the delivery systems required to carry nuclear weapons.

Earlier discussions dwelling on air war were (explicitly or just implicitly) skeptical about whether airborne nuclear retaliation would be sufficient to be unignorable and went ahead largely skipping mutual deterrence. Wohlstetter began and ended by establishing nuclear deterrence as the essential goal, but he reintroduced some explicit doubts about whether the punishments that could be inflicted in retaliation would be sufficient to the task.

Bernard Brodie published a collection of his studies at the RAND Corporation, in 1959, in a book entitled *Strategy in the Missile Age*,[20] which sold considerably more copies than had *The Absolute Weapon*. As in his earlier work, Brodie contemplated the hypothetical possibility that one side could escape nuclear retaliation by an assiduously designed counterforce attack, catching the other side's delivery systems on the ground. Responding to the sensitivity to the survivability of American bomber bases against a Soviet bomber attack, a concern that had made the rounds with Wohlstetter's team at RAND in the same years, and to a more widespread fear after Sputnik that the Soviets might inflict such a disarming first strike using a monopoly of missile capabilities, Brodie discussed these threats to deterrence at some greater length. Yet his conclusion was very much like that of 1946, that the retaliation a loser could inflict was likely to preclude the winner from becoming a total winner or from using nuclear weapons in the effort of winning.

As a later shot in this debate (whether the stress should be on countervalue impact, or on possible counterforce impacts blunting the countervalue impact), Brodie in 1978 questioned the attention Wohlstetter had received for the arguments of 1959, stating that "delicate" had probably never been an attribute of "the balance of terror."[21]

SOVIET ATTITUDES: THE NAGGING ISSUE OF ASYMMETRY

We turn now to a possibility of asymmetry that somewhat complicates the arguments between Brodie and Borden, or Brodie and Wohlstetter. Brodie (and NSC–68 and Kissinger) had concluded that only a few bombs in Soviet hands, if aimed at American cities, would be an unignorable deterrent, a factor making victory pointless once such bombs were used. Wohlstetter, in casting doubts about the assurance of destructiveness in nuclear confrontations, was really only questioning whether a few bombs in American hands, aimed at Soviet cities, would be as unignorable a deterrent for the Soviet leadership when it was contemplating victory. What if the Soviet leaders are more callous about the future of their urban centers than our leaders are about our cities? This is the possibility of asymmetry, which has been introduced often since the Soviet nuclear arsenal has grown, a possibility central to some of the scenarios of windows of vulnerability or Soviet superiority.

Wohlstetter was perhaps just as resigned as other strategic analysts to seeing the United States deterred by the consequences of nuclear war. But he was cautioning against any automatic assumption that Russia would be equally deterred. Much of the same line gets played out today in defenses of the Reagan administration's investments in counterforce strategic nuclear options.

A SIDETRIP: MCNAMARA, ANN ARBOR, EXTENDED DETERRENCE

The election of John F. Kennedy as President, with Robert S. McNamara assuming the post of Secretary of Defense, brought some major innovations to defense policy at the beginning of the 1960s, but these are not easily sorted with regard to the fundamental and enduring issues of nuclear strategy.

Kennedy had won office after directing two different criticisms at the Eisenhower administration in 1960. One Democratic message was that (because of the balancing of Soviet and American assured destruction forces) the United States would now be exposed to conventional, limited war attacks and would have to augment forces in this category to shield exposed allies in Western Europe and Southeast Asia (this basically being the argument already presented in NSC–68 and in Henry Kissinger's 1957 book). One could refer to the issue as the problem of *extended deterrence,* how one is to shield the promontories of the Eurasian continent against Soviet aggression, when a United States nuclear monopoly no longer exists to protect them by the prospect of nuclear retaliation against Moscow. The other criticism, in light of predictions of bomber gaps and missile gaps, was that the United States was falling behind in elementary strategic retaliatory power; the Soviets might even be able to escape retaliation altogether if they struck first, and the United States would soon encounter the danger of a failure of basic deterrence, never mind extended deterrence.[22]

McNamara and Kennedy quickly discovered that the second criticism was ill-founded. The Soviets had not produced large numbers of ICBMs after Sputnik, but rather had allowed the United States to generate a missile gap in reverse as production of Atlas, Titan, Minuteman, and Polaris missiles came along much faster than the worst-case American analysts had anticipated.

As the missile gap was dispelled in the new administration's first weeks in office, what did this suggest about the issue of extended deterrence? McNamara still believed that the most rational way to shield Europe against attack was to prepare for a conventional defense, that is, to induce NATO to match the United States in an investment in additional ground and tactical air forces. But what was the most rational approach to strategic nuclear forces, once the facts were digested that the United States was not in danger of a Soviet disarming counterforce, first-strike attack, but might almost be able to inflict such an attack on Russia instead?

West Europeans now showed signs of resisting the conventional rearmaments proposed by the Kennedy administration: they might be too costly economically and in the manpower committed to military service, and preparations for a nonnuclear defense of Western Europe might actually make the Soviet launching of a war more likely, when the threat of American nuclear escalation was lifted. Having denounced the Eisenhower administration for relying on nuclear escalatory threats as a way of extending deterrence to shield Europe as well as North America, McNamara and Kennedy now, after the Berlin blockade, found themselves moving into a repetition of the Eisenhower approach, cashing in on the surprising American advantage in intercontinental missiles. This was highlighted in Secretary McNamara's statement to the 1962 Athens NATO conference and in his subsequent Ann Arbor University of Michigan commencement speech, pronouncing a "no-cities" targeting policy by which (in the event of a Soviet conventional advance into Western Europe, where United States targets had not yet been attacked—the classical limited war nightmare) nuclear weapons would come into use against the Soviet Union, but only discretely and carefully aimed at Soviet military targets.

Perhaps United States nuclear weapons could indeed have been aimed with greater care. As part of the emphasis on being more rational than its Republican predecessors, the new administration looked to a more careful discrimination of military from civilian targets. Yet any analyst of the impact of nuclear weapons might still have concluded that substantial collateral countervalue damage would be inevitable in this kind of an attack. It was not the surgical cleanness of strike options that offered McNamara any new leverage here. Rather, what made the Ann Arbor threat more applicable was the windfall of surprising possibilities in any counterforce exchange that would follow.

Intent on winning the war (as before), intent on protecting the NATO allies to whom we had given our word (as before), feeling moral and rational about the countermilitary focus of our targeting (as before), intent on destroying Soviet strategic forces before they destroyed ours or destroyed our cities, United States strategic forces would now once again be brandished to dissuade Moscow from sending its tank columns into West Germany and beyond, even when those tank columns might not be repulsed by any of NATO's conventional weapons on the battlefield.

McNamara's explication of greater deliberation and rationality in nuclear targeting, however little difference it might have made for the number of Soviet citizens killed in an attack, had one immediate and important effect—it freed American nuclear target planners and weapons designers to find many more specific target aim points and thus to justify a far larger total of nuclear warheads for the United States stockpile.[23]

McNamara's commitment to the no-cities targeting doctrine, outlined at Ann Arbor, was to be quite transitory, as he soon became convinced that the degree of countervalue retaliation the Soviets could deliberately inflict (or the substantial collateral countervalue damage each side would inevitably inflict in the process of aiming at each other's military targets) would again be too much for the United States to bear.

The in-and-out Ann Arbor doctrine could thus in some ways be seen as supremely rational. Just as Bernard Brodie (or NSC–68) had outlined the changes in appropriate American strategy after the introduction of Soviet nuclear weapons, as compared with the period of United States monopoly, the United States would now have one appropriate response to a Soviet invasion of Western Europe when it had a counterforce potential very much greater than that of Russia (a potential for blunting all or most Soviet countercity retaliation), and another appropriate response if the Soviets reacquired an assured means of retaliation.

Analysts less committed to a forthright rationality might have feared, however, that the McNamara approach could strip away some possibilities of United States retaliation that had been maintained during the Eisenhower years, despite the alleged irrationality of any such escalation once the Russians had a nuclear force aimed at American cities. Anyone can threaten to escalate in defense of Europe if his own cities can escape the ensuing nuclear exchange; the difficulty in maintaining NATO deterrence comes in presenting a plausible threat of such escalation even if such cities are now vulnerable. McNamara's shift into, and then away from, his emphasis on counterforce, could have lead Moscow to expect that the United States would not dare to escalate once the Soviet strategic missile force had been augmented.

Many analysts would, of course, have questioned all along whether there was any reality to the Ann Arbor scenario. Too many Soviet citizens would have been killed in even the most careful United States counterforce strikes, and too many Americans would have been killed in Soviet last-gasp retaliation or in Soviet launches on warning, as Moscow felt itself in a position of "use them or lose them." Yet, as in earlier days, there was a political fact in the speeches of McNamara and Kennedy that could still have deterred the Soviets. As with Eisenhower and his policies, the Soviets could not be certain that Kennedy and McNamara did not actually foolishly believe in limited nuclear war, or rather in a primarily counterforce nuclear war. If American leaders believed in or were fixated on such nuclear war scenarios, it would be prudential for Soviet leaders to hold off on any exploitation of battlefield military advantages. Yet McNamara then stopped pretending to be fixated on such scenarios (some decades later even publishing accounts wholely disowning such scenarios).[24]

By the later 1960s, as America's attentions were monopolized neither by the defense of Europe nor by missile gaps, but rather by the commitments and frustrations of Vietnam, McNamara was ready to move toward a very explicit commitment to countervalue nuclear capabilities, on each side, now using the phrase "assured destruction" in his annual briefings on the defense posture of the United States. The phrase was waggishly lengthened by Donald Brennan to "mutual assured destruction," to make his point that the reliance on massive countervalue potential on each side could be seen as madness.[25]

Whatever was going to deter Soviet invasions of Western Europe now, it was not going to be any American counterforce capability offering a shield against Soviet nuclear strikes at the United States. With the European NATO members balking at going along with the McNamara proposals for an enhancement of conventional capabilities (and with U.S. Army forces in Western Europe themselves being thinned out to meet the demands of the war in Vietnam), it was also not going to be any clearly more reliable NATO conventional capability for a defense, any deterrence by denial instead of deterrence by punishment. Rather, it was probably the continued (and enhanced) deployment of theater nuclear weapons into the NATO Central Front, theater nuclear weapons that were likely to come into use even in any initially conventional war, if only because of the inherent risks of such use in the "fog of war" and "heat of battle."

Having dropped any explicit exploitation of counterforce nuclear capabilities, McNamara, for all of the reasons already noted, had hardly settled the United States into a total and simple acceptance of a countervalue balance of deterrents, a balance of terror. The additional nuclear weapons, which had begun to be procured with the initial McNamara venture into more militarily purposeful targeting, and the additional targets placed on the list did not get cancelled now, and the arsenal was destined to grow as if McNamara had never moved away from the Ann Arbor speech. The fundamental logic of that speech, that it was more rational and moral and purposeful to aim for a victory in the use of nuclear weapons (i.e., to aim counterforce at military targets, rather than simply to thrash around devastating the cities in which people live) continued to command lip service and intuitive respect.

CONTINUING OBSTACLES TO MORAL ACCEPTANCE

McNamara's first response to United States nuclear weapons targeting plans, that it was appalling that such weapons seemed to be aimed so mindlessly at civilian populations, is a normal first reaction, triggering moral revulsion and endorsements of nuclear or total disarmament in some people, and a return to a more traditional, careful, and high-minded counterforce stress on military targets in others. Building on some of the weapons systems and options launched in McNamara's tenure, Secretary of Defense James Schlesinger was to embark on a similar pattern in Richard Nixon's administration at the beginning of the 1970s in a pattern that Desmond Ball aptly labeled "déjà vu."[26] McNamara had shifted back to assured destruction in his own thinking and his public pronouncements, dismissing damage limitation and ABM deployments as technologically unfeasible, but the weapons programs that had commenced under a notion of more rational focus on military targets were not cancelled, and the dream of a more moral or more purposeful use of nuclear weapons was not ended. Perhaps it will never end. What Robert Jervis has referred to as the "conventionalizing of nuclear weapons" still has some appeal, at least until the inevitable collateral countervalue impact is once and for all accepted as a given.[27]

Will we ever see the day when most American scientists react with satisfaction, rather than with gasps of horror, to the degree of assured destruction in a nuclear exchange? Will we ever see the day when American clergymen can bring themselves to express satisfaction, rather than indignation, about this, because it makes a World War III so much less likely (even while it makes it so much more horrible)? Will we ever see the day when professional military officers take pride in how much countervalue damage they can inflict, and pride also in posing no counterforce threat to the adversary's ability to inflict such retaliatory countervalue damage on American cities? This would, of course, reverse all the professional imperatives they have inherited from centuries of Western tradition.

By the logic of mutual assured destruction, the logic that Brodie, Panofsky, and others have endorsed since 1946, it might now be time for such sentiments to begin winning some endorsement; but the philosophical, psychological, and sociological barriers to such endorsement are still substantial, with the public reaction to the findings of the nuclear winter study being a prime illustration.

SOVIET ATTITUDES AS A BARRIER

Will we ever see Soviet statements welcoming a pattern of mutual assured destruction? Soviet statements often stipulate that each side is now inevitably exposed to second-strike retaliation, but they do not go as far as Morgenstern or Panofsky or Brodie would have gone; they do not suggest that each side should *welcome* being vulnerable, that a country like Russia would take steps to leave itself vulnerable if it were ever thinkable that the other side's countervalue instruments might be blunted or laid open to a preemptive counterforce attack.

In their public statements, the Soviets, even more than the American community of theologians, scientists, and military professionals, have continued to indicate that the highest calling of any country's defense efforts is to shield its homeland against destructive enemy attack, including destructive nuclear attack. The best that the world has gotten from the Soviets, in terms of explicit statements, have been concessions that this may not be doable. Except for some important and interesting hints around the edges, however, we have not extracted any statements that this would be undesirable, even if it was doable, that this is something that a Soviet government official or defense planner should take steps to avoid.[28]

This Soviet obtuseness about the arguments for mutual deterrence and mutual assured destruction (this failure to accept the arguments outlined by Brodie, and tendency instead to sound more like an echo of Douhet or Borden) has, of course, nourished those Americans who continue to think of nuclear warheads as weapons, rather than simply as instruments of responsive retaliation. The feedback into the American strategic debate takes a number of forms, but it clearly has played a role, accounting in part for the change in tone from the Kennedy–McNamara years to the Reagan–Weinberger approach.

One form of such feedback is at the level of simple analysis of the technical facts. If the Soviets think that victory is meaningful in a nuclear war because the countervalue damage can be reduced enough to make it so (i.e., if the Soviets think that they should continue to strive to shield their civilian population against nuclear attack), perhaps they know something that we do not know and perhaps they are right.

Another form of feedback is more complicated. Even if we could never come to the same conclusion—that the countervalue damage of a nuclear war was bearable, so that winning was meaningful—we may have to contend with a Soviet leadership that has reached this conclusion; the Soviets may have settled into this view because of faulty analysis or because they are more callous than we are, much more concerned about political power and much less about civilian suffering. If so, according to an argument circulated widely in the Reagan administration, it is necessary for us to arrange for "victory" in a nuclear war to go to our side rather than to the Soviet side, not because such a victory would be meaningful for us, but because this would be the only sure way to deter Moscow (because victory and political power are indeed meaningful to them). The Politburo can be counted on to avoid launching a nuclear war that it would lose, so the argument goes; it could not be counted on to avoid launching a nuclear war that merely brought nuclear countervalue retaliation against Soviet cities, if Moscow, in the meantime, saw prospects of "winning" that war in the grand counterforce exchange.

The idea that the Russians are different from Americans, different particularly in not regarding atomic or hydrogen bombs as the absolute weapons, has thus played a major role in stoking renewals of the argument. The proponents of counterforce applications of nuclear weapons might finally give up if the countervalue impact was, once and for all, stipulated to be so ignorable that military outcomes would become a subject of indifference; but this stipulation may have to come more globally, not just from North America, if it is to terminate effectively a debate that has already gone on too long.

AN ABSTRACT COMPARISON

One could illustrate some or all of the unresolved dilemmas here by trying to lay out and evaluate the alternative possibilities shown in the following matrices. We can begin by contemplating a

STABLE

Counterforce Capabilities:	U.S. > USSR	U.S. = USSR	U.S. < USSR
U.S.	CF	—	—
USSR	—	—	CF

Figure 3.1

military technology in which neither side can escape the other side's retaliation (i.e., in which neither side has a meaningful counterforce capability), unless it tremendously outnumbers the nuclear forces of the other side. Such probably was the case with the confrontation of single-warhead missiles after the middle of the 1960s, as the poor accuracies of these missiles would have required that more than one ICBM be expended for each ICBM dug out of its concrete silo on the other side. Along a continuum of ratios of Soviet to American total forces strengths, we thus denote the existence or absence of this meaningful counterforce capability. See Figure 3.1.

We would compare this then with the technology of bomber forces facing bomber forces earlier, or a situation later of multiple-warhead missile forces facing each other (omitting, for the sake of logical illustration, the beneficial complication of the existence of submarine-based nuclear missile forces). See Figure 3.2.

UNSTABLE

Counterforce Capabilities:	U.S. > USSR	U.S. = USSR	U.S. < USSR
U.S.	CF	CF	—
USSR	—	CF	CF

Figure 3.2

STABLE

Countervalue Capabilities:	U.S. > USSR	U.S. = USSR	U.S. < USSR
U.S.	CV	CV	—
USSR	—	CV	CV

Figure 3.3

If one wished to use a different notation to depict exactly the same array of possibilities, we could show the absence or presence of retaliatory second-strike countervalue capabilities, first dealing only with secure and hardened single-warhead missiles. See Figure 3.3.

Next, we can deal with less stable confrontations of bombers versus bombers, or MIRV missiles versus MIRV missiles. See Figure 3.4.

Combining our notations, we should note that we have essentially four distinct situations, as shown in Figure 3.5.

As a way of sorting out much of the current strategic debate, and other earlier strategic debates, we might ask someone to designate which was his or her first choice, and which was his or her last, of the four bundles above.

A believer in MAD might list a first choice as *situation grey*, in which each side would have secure second-strike retaliatory ca-

UNSTABLE

Countervalue Capabilities:	U.S. > USSR	U.S. = USSR	U.S. < USSR
U.S.	CV	—	—
USSR	—	—	CV

Figure 3.4

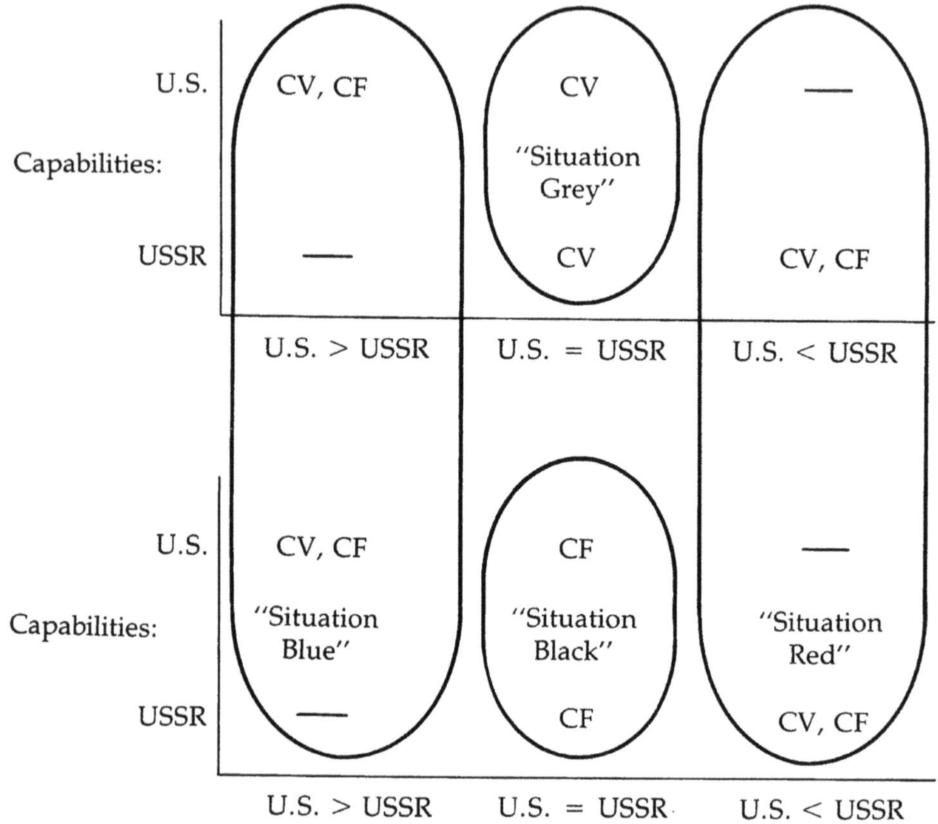

Figure 3.5

pabilities, and he might list his worst nightmare as *situation black*, the unstable confrontation in which either side might hope to escape such retaliation—if only it could beat the other side in an elegant counterforce sneak attack (all the while fearing that the adversary was contemplating just such a counterforce preemptive attack itself, because the capabilities for such a productive use of nuclear weapons would be plausible in either direction). As illustrated by the ideas we have cited of Brodie and others (perhaps even of Wohlstetter in his 1959 article), the ranking would be as follows:

Grey
Blue
Red
Black

By contrast, a supporter of the policies of the Reagan administration might list his or her happiest choice as *situation blue*, in which the United States could hope to escape such retaliation (but the USSR surely could not), and a worst nightmare would be the reverse case, *situation red*, in which the Soviets would have prospects of carrying through such a splendid first strike, but the United States would not. Here, the ranking would run as follows:

Blue
Grey
Black
Red

Both schools of American strategic thought would agree on preferring American superiority to Soviet superiority (blue over red), and on preferring a stable parity to an unstable parity (grey over black). The schools would differ very much, however, over whether blue or grey would be the very best case, and whether red or black would be the very worst.

How could supporters of the Reagan administration or any other Americans expect the Soviets to trust the United States with the advantages suggested in situation blue? We Americans might feel we proved our peaceful nature and good will when we alone had nuclear weapons in the immediate aftermath of World War II (that indeed was situation blue). Perhaps, as Bernard Brodie argued in his 1946 book, the United States would never be able to initiate a war, but would only respond if first attacked by an aggressor, as at Pearl Harbor.[29] Because the Soviets would not be so foolish as to supply this pretext until they had nuclear weapons of their own, Brodie felt it quite appropriate to devote most of his analysis to the situation in which both sides had such weapons.

Yet perhaps the United States was not simply held back by high-mindedness after 1945. Skeptics would point to our unilateral use of nuclear weapons on Hiroshima and Nagasaki (with the allegation sometimes being advanced that this was done more to intimidate the Russians than to defeat the allegedly already defeated Japanese), and to a host of other realistic power politics factors that made the United States reluctant to use atomic bombs when we alone had them, not the least being the easy advance Soviet ground forces might have made into Western Europe during and after a United States nuclear air attack. Would we have dropped nuclear weapons on the Soviet forces garrisoning Bruxelles and Paris? The aforementioned doubts about the absolute power of fission bombs (as opposed to the later fusion thermonuclear H-bombs) also may have played some reasonable role in discouraging American leaders from trying to apply and exploit and perpetuate the United States monopoly. Additionally, we have learned more recently that not very many nuclear weapons were produced for the American stockpile in 1946 and 1947; the stockpile in many ways amounted to a myth until after the Soviets had shown themselves to be accumulating such a stockpile.

Even if the United States had somehow "proved itself" in the 1945–48 period, moreover, would we be as restrained in the future if we could ever recover such an advantage? Or, might we not this time exploit such a monopoly, just as we have always feared that the Soviets would exploit a monopoly on their side, using our again unique threat of countercity nuclear attacks to force a surrender, to force a perpetuation of the monopoly we had revived. And if the Soviets have reason to trust us on the basis of the 1945–48 experience, can we count on them to remember these reasons and not to panic about the end of their countervalue retaliatory hold on the cities of the United States?

Similarly, how could some right-wing Americans prefer the seeming, but highly unstable and mutually nerve-wracking, equality of situation black over the apparent Soviet monopoly of situation red? There is no denying that either is a bad situation from an American point of view, and serious efforts must be expended to keep out of these boxes. Yet the seeming advantage of black over red may be illusory or worse, for the *mutual* capabilities for preemptive attack could drive Moscow to fire off missiles, which it otherwise would not have fired, in a war nobody wanted.

Clearly, the desirable thrust, in avoiding the unilateral Soviet advantages we have marked as situation red, is to steer toward situation grey, in which neither side would be inclined to strike first in a crisis, even if military victory was the issue. (Taking all of human values into account, neither side would ever look forward to being struck; but if victory was the only consideration in such a technological confrontation, it would actually be best for either side if the other side made the mistake of striking first, using up more of their strategic forces than they destroyed of ours—a very desirable situation for the sake of crisis stability.)

The analysis of Oscar Morgenstern in 1959 was very explicit in indicating a preference for our situation grey over our situation blue. The earlier 1946 analysis of Bernard Brodie, Brodie's book of 1959, and the 1973 Panofsky response to Fred Ikle's[30] questioning of the entire logic of deterrence all indicate such a preference.

Yet there are miles to go before the last American that matters on such questions will have accommodated himself to this judgment.

When the only explanation for a failure to move to a first-strike capacity, or an SDI–ABM system, is its alleged technological infeasibility, we have not yet made the necessary breakthroughs in the dissemination of deterrence logic.

Much the same has to be said, of course, about Soviet public statements, which may at times concede an unavoidability of the situation we have coded a stable and desirable grey here, but which do not yet stipulate that this would be preferable to our situation red, whenever red were physically attainable. It is Moscow's preference for red over grey that fuels a considerable part of some Americans' preference for blue to our situation grey, and even a more frightening background preference for our situation black over red.

The aim of the USSR toward situation red and of the United States toward situation blue together tend generally toward the worst of all situations, our situation black, in a flow of hard-target accuracies, multiple warheads, and war-fighting communications satellite systems that might one day plunge us into a war nobody wanted.

SOME CONCLUSIONS

We have guessed that our two major American perspectives on the strategic debate might at least agree that the situation grey, in which no one had a counterforce capability, was preferable to situation black, in which both superpowers had it. Peace would be less threatened if neither side saw any point in striking first during a crisis, if each simply held back to use its nuclear forces in retaliation after the other had used them first.

When the options are outlined abstractly as in the matrices in Figures 3.1–3.5, this much agreement might at least be achieved. When they are discussed in more normal terms, however, all the blurring we have already noted tends to return, as the traditional morality of aiming at weapons rather than at people reasserts itself, and as Secretaries of Defense agonize in public about wanting something else to do with nuclear weapons, besides using them to murder millions of innocent people, or not using them at all.

It is thus bizarrely possible that the American Roman Catholic Bishops' statement questioning countervalue targeting,[31] or the policy suggestions appended to public presentations of the nuclear winter study,[32] will also develop a momentum toward the unstable black situation ahead of the stable grey. This is part and parcel of the tragedy of the unfinished debate we have been discussing here, a debate that was launched, if not earlier then immediately after Hiroshima, in the writings of Brodie and Borden.

Perhaps the greatest failing in the acceptance of post–World-War II deterrence theory would come here, in such comparisons where Americans and others might subliminally welcome rather than reject counterforce capabilities, where moral theorists and others might support a targeting of opposing military forces, rather than simple murderous retaliation against innocent civilians. President Reagan's Star Wars announcement of March 23, 1983, in which he endorsed proposals for a new ABM system so that future generations would supposedly be shielded rather than aimed at, may have been a politically cynical exploitation of the gut sentiments of many Americans here, a move to deflect any moral criticisms of American nuclear weapons procurements, as exemplified in the statement of the Roman Catholic Bishops. Yet it showed the strength of a tradition of decades and centuries, a tradition in all of Western culture, which was indeed particularly strong in the military doctrine of the United States through two World Wars, that civilians should not be aimed at directly as a target, that any punishment of civilians should instead occur only as the unavoidable byproduct of attacks on legitimate military targets.

The attitudes we have discussed here are bad news for world stability and for the solving of the international security dilemma; they are also evidence that the logic of deterrence has not been as persuasive, or as widely understood, as many of us had assumed.

It will be difficult to tell whether we were ever as close to winning the wider acceptance of mutual assured destruction as we might have hoped, the acceptance that Donald Brennan was lamenting as he unsuccessfully argued for a commitment to ABM at the end of the 1960s. It will also be difficult to tell whether American attitudes might not have settled into a greater commitment to deterrence and strategic stability, if Soviet public pronouncements had not remained so committed to the endorsement of military advantages and counterforce military plans, wherever these might be possible.

But one cannot, in any event, blame deterrence theory for its alleged intellectual failings, or for a failure to solve this decade's international security problems, if such theory has not yet been understood or accepted by many Americans, Russians, Europeans, or others.

NOTES

1. An example of such criticism from the left can be found in Philip Green, *Deadly Logic* (Columbus: Ohio State University Press, 1966).

2. As a right-wing criticism, see Colin Gray, "Nuclear Strategy: The Case for a Theory of Victory," *International Security* vol. 4, no. 1 (Summer 1979) pp. 54–87.

3. Bernard Brodie, ed., *The Absolute Weapon* (New York: Harcourt Brace, 1946).

4. William L. Borden, *There Will Be No Time* (New York: Macmillan, 1946).

5. Giulio Douhet, *The Command of the Air* (New York: Coward-McCann, 1941); and J. F. C. Fuller, *The Reformation of War* (New York: E. P. Dutton, 1923).

6. On bureaucratic politics theories explaining the evolution of strategic pronouncements and weapons procurements, see Morton H. Halperin, *Bureaucratic Politics and Foreign Policy* (Washington, D.C.: Brookings, 1974).

7. On the reactions of the Bikini test, see Gregg Herken, *The Winning Weapon* (New York: Alfred A. Knopf, 1980), pp. 224–26.

8. The text of NSC–68 can be found in *Naval War College Review* vol. 27, no. 6 (May-June 1975), pp. 51–158.

9. Henry A. Kissinger, *Nuclear Weapons and Foreign Policy* (New York: Harper and Row, 1957).

10. Henry A. Kissinger, *The Necessity for Choice* (New York: Harper, 1961).

11. Henry A. Kissinger, *The Troubled Partnership* (New York: McGraw-Hill, 1965).

12. The Kissinger Moscow quote is cited in Edward Zuckerman, *The Day after World War III* (New York: Viking Press, 1984, p. 261.

13. Henry A. Kissinger, "Strategy and American Alliance," *Survival* vol. XXIV, no. 5 (September-October 1982), pp. 194–200.

14. McGeorge Bundy, "To Cap the Volcano," *Foreign Affairs* vol. 48, no. 1 (October 1969), pp. 1–20; and Wolfgang Panofsky, "The Mutual Hostage Relationship between America and Russia," *Foreign Affairs* vol. 52, no. 1 (October 1973), pp. 109–18.

15. For a very well researched example, see David A. Rosenberg, "A Smoking Radiating Ruin at the End of Two Hours," *International Security* vol. 6, no. 3 (Winter 1981–82), pp. 3–38.

16. Albert Wohlstetter, "The Delicate Balance of Terror," *Foreign Affairs* vol. 37, no. 2 (January 1959), pp. 211–34.

17. Albert Wohlstetter, "Nuclear Sharing: NATO and the N + 1 Country," *Foreign Affairs* vol. 39, no. 3 (April 1961), pp. 355–87.

18. Wohlstetter, "The Delicate Balance of Terror," p. 229.

19. Oscar Morgenstern, *The Question of National Defense* (New York: Random House, 1959).

20. Bernard Brodie, *Strategy in the Missile Age* (Princeton, N.J., Princeton University Press, 1959).

21. Bernard Brodie, "The Development of Nuclear Strategy," *International Security* vol. 2, no. 4 (Spring 1978), p. 69.

22. For a good overview of the various thrusts of policy change under McNamara, see William W. Kaufmann, *The McNamara Strategy* (New York: Harper and Row, 1964).

23. See Desmond Ball, *Politics and Force Levels* (Berkeley: University of California Press, 1980).

24. Robert S. McNamara, "The Military Role of Nuclear Weapons," *Foreign Affairs* vol. 62, no. 1 (Fall 1983), pp. 59–80.

25. See Donald J. Brennan, "The Case for Population Defense," in *Why ABM?*, ed. Johan Holst and William Schneider (New York: Pergamon, 1969).

26. Desmond Ball, *Deja Vu: The Return to Counterforce in the Nixon Administration* (Los Angeles: California Seminar on Arms Control and Foreign Policy, 1974).

27. See Robert Jervis, *The Illogic of American Nuclear Strategy* (Ithaca, N.Y.: Cornell University Press, 1984).

28. On Soviet statements and actual feelings, see Francis P. Hoeber and Amoretta M. Hoeber, "The Soviet View of Deterrence: Who Whom?" *Survey* vol. 25, no. 2 (Spring 1980), pp. 17–24; and Raymond L. Garthoff, "Mutual Deterrence and Strategic Arms Limitation in Soviet Policy," *International Security* vol. 3, no. 1 (Summer 1978), pp. 112–47.

29. Brodie, *The Absolute Weapon*, pp. 87–8.

30. Fred C. Ikle, "Can Nuclear Deterrence Last Out the Century?" *Foreign Affairs* vol. 51, no. 2 (January 1973) pp. 267–85.

31. National Conference of Catholic Bishops, *The Challange of Peace: God's Promise and Our Response* (Washington, D.C.: United States Catholic Conference, 1983).

32. On nuclear winter, see Carl Sagan, "Nuclear War and Climatic Catastrophe," *Foreign Affairs* vol. 62, no. 2 (Winter 1983–84), pp. 257–92.

CHAPTER 4

On the Tiger's Back: The Development of the Concept of Escalation

LAWRENCE FREEDMAN

Once on the tiger's back we cannot be sure of picking the place to dismount.

George Ball, October 1964

INTRODUCTION

Escalation has become one of the most influential of the family of concepts developed by the strategic studies community in the late 1950s and early 1960s. It has passed into everyday language to the extent that it is now almost impossible to discuss the course of any conflict, industrial or political as well as military, without some reference to escalation. Yet it has reached its current prominence without the benefit of a precise definition.

Students of strategy generally agree that escalation refers to an increase in the scope and intensity of a given conflict. But for some students, it captures a phenomenon to be observed at work in all wars, tending to move them along independently of the wishes of the participants. For others, escalation describes a deliberate act, a conscious commitment to raise the level of fighting, usually in the belief that this act will be understood in the appropriate manner by the enemy and its consequences thereby controlled. For some, the basic issue captured by the idea of escalation revolves around the ultimate destination—possibly an all-out nuclear war; others are concerned with exploiting the sense of movement. The movement itself

The author is grateful for valuable comments on an early draft of this paper from Dr. Barry Buzan, Professor Alexander George, Professor Robert Johnson, Professor Barry Steiner, Dr. James Thomson, Dr. Philip Sabin, and Dr. Richard Smoke.

can be described in terms of a steady, unbroken growth or a series of discrete steps—the *escalation ladder*.

The first serious academic analysis of the concept of escalation was published in 1977, long after the term had entered the vernacular. In this study, Richard Smoke observed that the term had acquired many meanings. Smoke himself came down in favor of a definition of escalation as an

> action that crosses a saliency which defines the current limits of a war, and that occurs in a context in which the actor cannot know the full consequences of his action.[1]

Smoke's careful formulation helps us obtain the maximum amount of analytical value from the concept. It carries a warning found in many studies of the subject: that escalation is more difficult to control than those responsible for its initiation might think. However, this was not the original way in which this term was used. Smoke's definition is a pessimistic variant of a view that saw escalation as the deliberate crossing of a saliency, which could serve as an effective method of putting pressure on an adversary to settle on favorable terms. This optimistic view had displaced an earlier and more pessimistic view of escalation as an essentially uncontrollable phenomenon liable to turn a limited war into a general war.

It is not the objective of this article to choose between the alternative concepts or to develop and refine one particular view of escalation. The objective is to describe the origins of the concept and its treatment in strategic studies literature and in United States crisis diplomacy. It illuminates the interaction between the development of nuclear strategy and the practicalities of crisis and conflict in Cuba and Vietnam, as well as the interaction between policy makers and the leading civilian defense analysts, then at the height of their powers and influence.

FROM PESSIMISM TO OPTIMISM

Smoke suggests that the confusion to which he draws attention is the result of the speed with which escalation captured the "verbal terrain." This caused the meaning to be enlarged and so become more vague.[2] The growth in its use was certainly rapid. It was not used publicly in a strategic context until as late as 1959 and then by a British advocate of disarmament. By 1961, it was well established. In that year, for example, President Kennedy referred in a budget message to Congress, in a usage that was by then standard, to the threat of "the escalation of a small war into a large one."[3] By the mid-1960s, commentary on the Cuban missile crisis and the growing American involvement in Vietnam, as well as two important works of strategic analysis with escalation in their titles (by Herman Kahn and Bernard Brodie), had popularized the term. Fred Ikle, reviewing those two books in 1967, noted that the many ways in which limited wars became less then limited had become:

> tucked under the label "escalation," a word scarcely fifteen years old (in its present meaning), which we find daily in our newspapers though not yet in most of our dictionaries.[4]

Escalation was soon used in quite different and incompatible ways. As early as 1963, Morton Halperin was drawing attention to the need to distinguish between different types of activity that were being obscured by the catchall term, but his own alternative terminology failed to take hold, and references to escalation became even more frequent.[5] Certainly, apart from Herman Kahn, there was little exploration of the concept.[6] Bernard Brodie wrote the other important work with escalation in its title,[7] but he was preoccupied in this book with making a point, not elaborating a concept. After Brodie, the most substantial study of escalation, undertaken at RAND, still failed to provide a clear definition of the concept. Perhaps by this time the term had become so general that no need was seen for definition.[8] As we have noted, not until Richard Smoke's book in 1977 was there a serious attempt by an academic to unpackage the concept.[9]

However, this analytical confusion was not simply a function of the speed of entry into the language. It was the result of a direct challenge mounted by leading figures in the strategic studies community against what they considered was a misleading metaphor that overstated the problems facing attempts to control conflicts.

Escalation was originally adopted as a relatively straightforward metaphor, describing the pressures that could transform a limited war into something more widespread and destructive. It was initially used in critiques of the versions of limited war theory that allowed for use of tactical nuclear weapons on the battlefield. It is still often used in this way.

Scholars who argued that this common use exaggerated the difficulty both of establishing and sustaining limits on war and of moving beyond them in a controlled fashion, initially sought to qualify the original metaphor. However, within a couple of years of its introduction, the word *escalation* began to be redefined by the strategists to refer to the means by which there could be a controlled movement through the stages of a conflict, so that the intensification of the scope and level of warfare did not necessarily involve a loss of control or the abandonment of political objectives. In 1967, reflecting this more positive approach, Iklé observed: "Escalation makes us think of step-wise increments, each of which confronts the enemy with noticeable challenge."[10] John Garnett commented in 1970 that having been regarded as a "tragedy, a disastrous, uncontrolled extension of the scope and violence of war," escalation had become a "*strategy*, a deliberate instrument of state policy."[11]

The concept of escalation attained its most optimistic and positive formulation at the peak of United States confidence in its capacity to manage the pressures of conflict, crisis, and war. Thereafter, the experience of its supposed application in Vietnam led to a derogatory connotation: escalation might be deliberate—but it was probably dangerous and unwise.

THE PROBLEM OF LIMITED WAR

The sort of tendencies referred to as escalation have long been familiar to students of war. However, their range, diversity, and differential impact have precluded their confinement in a single strategic category. The difficulties of sustaining any limitations in warfare had long been recognized, as had the extent to which these might be the result of one side upping the ante and the other following. Clausewitz's notion of "reciprocal action" is a good example of this.[12] Even in the contemporary period, the basic ideas explained by reference to escalation were well established before the term was coined.

The stimulus to innovation in strategic thought during the 1950s was the developing nuclear stalemate that appeared to rule out general war as a way of resolving disputes among the great powers. The initial response was to argue that as one could no longer have all-out war, then it would be impossible to pursue all-out objectives. That assumed that the objectives set the tone for the conduct of war. Unfortunately, the objectives with which belligerents start wars rarely stay fixed during its course. Furthermore, the means adopted by one side must be determined as much by those adopted by the other, as by a sense of proportion with objectives.[13]

Some people toyed in the mid-1950s with the idea that limited use of nuclear weapons *could* be proportionate to limited political objectives. Either the so-called small or tactical nuclear forces could be used to make up for deficiencies in conventional firepower and so bolster the defense, or else it would be possible to move forward with graduated steps, steadily increasing the pressure on the enemy to cease his aggression. Much of the literature of the second half of the

1950s was preoccupied with these propositions.[14] One example of this is the idea of *graduated deterrence*, popularized in Britain by Rear Admiral Anthony Buzzard. This approach depended on the assumption that there were distinct stages in war that would be well understood. The threat to move from one stage to the other would serve as a powerful deterrent.[15]

Critics suggested that limits would be difficult to observe, even if, at first, only tactical nuclear weapons were used, and that the belligerents could soon find themselves engaged in all-out nuclear war. This suggested to some analysts, whose numbers and influence grew during the 1950s, that limited wars were best fought with conventional weapons, and that nuclear weapons would have little strategic role other than the crucial one of deterring nuclear use by the other side. Others, however, argued that the disproportionate nature of nuclear use could be exploited so that evidence that one side was prepared to risk nuclear war (brinkmanship) might alarm the other side and encourage a more accommodating position.

As we have noted, the term *escalation* was first used to refer to the process by which a limited war could get out of hand. The process was referred to very often in the debates of the 1950s[16] but in a different language, using such words as *explosion* and *expansion*, along with *trigger*, *spiral*, *eruption*, and *spread*. For some time, they were used interchangeably with *escalation*. In addition, even the more elaborate ideas later associated with *escalation* can be traced to the 1950s. As we have noted, the idea of proceeding along graduated stages can be found in the work of the British graduated deterrence school.[17]

Meanwhile, the theorist whose name became closely associated with the concept was developing his main ideas during this period. Thomas Schelling is regularly classed with Kahn as one of the major theorists of escalation, and, as Smoke has shown, his ideas are critical to its development as a serious concept. Kahn also expressed his debt, suggesting that Schelling's notion of a "competition in risk-taking" was the starting point for his own theories.[18] Yet even during the 1960s, Schelling used the term only sparingly. In his seminal *Strategy of Conflict*, published in 1960, it does not appear at all, though most of his major theoretical insights are contained in this book. For example, the chapter entitled "The Threat Which Leaves Something to Chance" is a classic discussion of the ways in which a war might get out of control and the extent to which this process might be exploited for strategic advantage. "The idea," he writes, "is simply that a limited war can get out of hand by degrees."

A variety of actions might increase the sensation of being out of control. The hope is that by sharing this sensation with the enemy, he is provided with "an overpowering incentive to lay off." The image of war conveyed by Schelling is of something gradually getting out of control until it becomes total and all-encompassing. His strategic insights concentrate on the extent to which the early stages of this process can be controlled and skillfully exploited to compel the other side to agree to acceptable terms for a settlement. He also warns of the danger of events rapidly getting out of control because of the fears of belligerents: "People may vaguely feel they perceive that the situation is inherently explosive, and respond by exploding."[19]

Here we see the origins of a tension that perplexed governments and analysts during the following decade. The United States was committed to using nuclear weapons, should its allies be subject to aggression, and there was a widespread expectation that nuclear weapons would be used as a matter of course at the start of any major conflict. Yet any use would only make sense if it could be controlled, especially when it came to limiting retaliation against American territory. But if a method of nuclear use that limited the risks of devastating retaliation could be found, might it not be as available to the Soviet Union as to the United States?

Out of this concern came two dominant themes in strategic studies: the possibility of establishing limits on the employment of nuclear weapons in a major war; and the premium that might be placed on a preemptive strike stemming from the degree of vulnerability of strategic forces to a surprise attack.

Raising these issues indicated a declining confidence in the maintenance of limitations. Thus, when a collection of knowledgeable persons was asked to rank the sources of risk of accidental war (defined loosely as war by miscalculation or unintentional war), those who did so put the spread of limited war as the most serious danger. The study went on to observe: "any kind of armed conflict on a major scale in Europe which is not rapidly brought under control would probably trigger a general nuclear war."[20]

THE TRAGEDY OF ESCALATION

In these debates, the word *escalation* was soon in service. It was valuable as a compelling metaphor. Unlike those terms drawn from the stark vocabulary of military planning (*worst-case scenario, preemptive strike, overkill*) that have since made inroads into popular speech, *escalation* involved an imagery that was readily accessible and comprehensible. Moving staircases were first introduced around the turn of the century,[21] and the idea of the escalator soon found a place in fiction. It was even used at least once in the 1930s to describe an arms race.[22] Also in the 1930s, escalator clauses first appeared in contracts to provide for periodic upward or downward adjustments.

Moving staircases served as a natural source of metaphor. The first use of *escalation* in its contemporary context (at least according to the *Supplement to the Oxford English Dictionary*, as well as my own research) was in 1959 by the English writer, Wayland Young, then an active proponent of nuclear disarmament. It came in an argument against strategies of limited nuclear war and graduated deterrence based on the small nuclear weapons then coming into the inventories of the Western powers.[23] Young stated that:

> The main argument against this policy is the danger of what strategists call escalation, the danger that the size of the weapons used would mount up and up in retaliation until civilization is destroyed as surely as it would have been by an initial exchange of thermonuclear weapons.

In the glossary, we find the following definition:

> Escalation–Escalator: The controlled exchange of ever larger weapons in war, leading to the destruction of civilization.[24]

At least one reviewer picked up the term. In *Survival*, the journal of the newly formed Institute for Strategic Studies in London, the historian Noble Frankland noted: "There is in war an inherent tendency towards what in the barbarous jargon of the day is called 'escalation.'"[25] As with Young, there is an implication that the term originated elsewhere—among the strategic studies community—but as yet no earlier published reference has been found. Young (now Lord Kennet) suggests that his source might have been Henry Kissinger, with whom he was in touch at the time.[26]

Not long after the book's publication, reference appeared in the liberal press. In December 1959, the distinguished physicist and commentator on nuclear matters, Professor Patrick Blackett, made passing reference (in the course of a discussion of the dangers of tactical nuclear warfare) to "the risk of starting the process of escalation towards total war."[27]

Around 1960, a use of *escalation* along these lines began to appear in the mainstream American literature, and it moved soon afterward into popular use.[28] The term, therefore, was not introduced into the strategic debate so much as a developed concept (unlike, for example, the concepts of first and second strike), but as an extremely convenient metaphor requiring little explanation. The essential idea of the escalator was of an upward movement that could not be stopped.

Once one had committed oneself to the moving staircase one had to move upward (or downward!), however much the original decision to step on it might be later resented or regretted. Escalation, therefore, was not so much a strategy as it was something to be avoided: a challenge to the concept of limitations but not yet a means of controlling the easing of these limitations.

In the United States, the term was initially used the same way as in Britain—by way of reference to a process to be avoided. It was noted earlier that it might have been Henry Kissinger who transmitted the term to Wayland Young in Britain in 1959. He certainly used it himself in 1960 in his *Necessity for Choice*. This is not surprising as, if nothing else, Kissinger was always acutely sensitive to changing trends in strategic studies. The context is made clear by the index reference: "Escalation, risk of in limited war." In the text, escalation is defined as: "The addition of increments of power until limited war insensibly merges into all-out war." This is presented as a problem with theories of limited nuclear war, such as his own of a few years earlier (which was being reassessed in this essay). States may prepare to engage in a limited nuclear war as an alternative to total war, but once embarked on a limited war they have made the total war inevitable. Although he considered the argument overstated, it was also:

> irresponsible to minimize the risk of escalation. It will be great in proportion as limited war is conceived entirely as a strategic problem rather than as an opportunity for negotiations.

Elsewhere, he stated that it would be a product of the nature of the limitations and the duration of a conflict, so that a "limited nuclear war lasting one day may involve a smaller danger of escalation than a conventional war lasting a year."

Escalation, therefore, was something that was risked as soon as war began, and the risk grew with time and with the level at which it was attempted to sustain limitations. Kissinger's argument was that by combining military action with negotiations, the risk could be alleviated. Thus, escalation was not an aspect of what later came to be described as coercive diplomacy.[29] Rather, it was what happened if war was conducted in the absence of effective diplomacy.

There was, however, a brief acknowledgment of the point later elaborated by Schelling and others: a key aspect of the bargaining process in war might be the respective assessments of each other's willingness to risk all-out war. If the West was unwilling to accept this risk, then "the aggressor may be encouraged to outwait us or to engage in a test of nerves with a built-in escalator to all-out war."[30]

Another early use of the term is found in Herman Kahn's *On*

Thermonuclear War, published in 1960. Here, *escalation* was used as if it were already a reasonably familiar term (but Kahn tends to do this).[31] The use was similar to that found elsewhere at the start of the decade and quite at variance with Kahn's own later use. Consider the only passage in which he actually defined the term:

> Nearly as worrisome as the war by accident is the war by miscalculation. The most likely way that this would occur is by escalation—the unpremeditated increase or spread of a limited operation.

Escalation was a "threat," and, as such, "an important deterrent to limited actions." If "reliable measures to limit escalation" could be introduced, this would "inevitably increase the probability of the use of limited levels of violence." Nevertheless, it was "still important to control or limit the possibility of escalation even if doing so decreases deterrence."[32]

Elsewhere, he addressed the argument that a "residual fear of war" in a seemingly stable balance of terror would "deter the other from starting a process that could lead to escalation." Here he did suggest that escalation could be a deliberate act, but also that it would be misguided.[33] He also noted the extent to which a threat of reprisal or escalation might help obtain a negotiated solution to a conflict.[34]

Glenn Snyder's *Deterrence and Defense* included a similar use: escalation described the upward thrust to higher levels of destruction that could occur in any war, recognition of which might well govern key strategic choices. Thus, Snyder made most use of the term in a paragraph noting how, if such escalation came to be seriously feared during a limited conventional war, prewar calculations concerning the inadvisability of a nuclear first strike might be cast in a different light. If "high levels of destruction" were virtually inevitable, one side "might prefer the consequences of striking first all-out to those of 'riding the escalator.' " Although Snyder may have found the metaphor convenient (and as can be seen from the above quote he was using it quite precisely), he was by no means dependent on it. Elsewhere, he referred to the "inadvertent explosion of full-scale war" and "tactical nuclear warfare exploding into full-scale nuclear conflict".[35]

THE FIREBREAK

References to escalation in the early months of the Kennedy administration were very much of this form. Escalation appeared as a threat to be avoided, a challenge to an administration convinced that it must remain at all times master of events. In Robert McNamara's Department of Defense, the emphasis was on strict control of all activities, from budgeting to weapons design. Escalation, in the sense of being driven along by the pressure of events to a disastrous conclusion, was utterly antithetical. All moves should be deliberate and calculated, preceded by a full assessment of available options. The conduct of war, whether conventional or nuclear, was not excluded from this requirement. The desire to ensure that there was no unwarranted escalation was evidenced by the measures taken to improve command, control, and communications, the concern with the temptations to preempt should retaliatory forces be left vulnerable, the drive to increase nonnuclear options to avoid reliance on incredible nuclear threats, and the exploration of controlled nuclear options for the worst contingencies.

William Kaufmann, writing in 1964 as a publicist for *The McNamara Strategy*, consistently described escalation as a strategy to be avoided: the objective of McNamara's approach was to find ways to "escape from" becoming a "prisoner" of such a strategy. Unfortunately, a strategy based on the threat of escalation had found favor in Europe:

> a series of conventional and nuclear tripwires linked like firecrackers, which would explode with automatic and increasing fury if the Soviets tried to press through them.

The Soviet strength in conventional arms would place it under no pressure to withdraw at an early stage of the conflict, thus forcing the United States "to climb inexorably up the ladder of escalation." This was "too automatic an approach to McNamara."[36]

When administration spokesmen referred to escalation, it was by and large as the process that would get underway once the first nuclear weapons were used, and Kaufmann certainly follows this use. Thus, just as escalation could well take over once the first nuclear weapons were used, before that point there could be a controlled and flexible response. To explain policy, administration spokesmen referred to the need to cope with Soviet aggression mounted at every level along the "spectrum of force." Kaufmann claimed that:

> The non-nuclear build-up will increase our capacity to tailor our responses to a particular military challenge to that level of force which is both appropriate to the issue involved and militarily favorable to our side.[37]

These concerns were expressed in the language of contemporary strategic analysis. In late 1962, John McNaughton, a senior

Pentagon official with special responsibilities for arms control and a friend of Schelling's, noted "a tendency for every confrontation to spiral upward in violence." He spoke of the mutual desire of the United States and Russia to avoid "the eruption, escalation or prolongation of nuclear war by accident, or miscalculation," and of the value of "building firebreaks against escalation of conflict."[38]

The idea of the "firebreak" was central to this approach. This was brought out in February 1963 in a speech by Alain Enthoven, then Deputy Assistant Secretary of Defense (Systems Analysis). Enthoven described a policy objective of enabling the president to be able to choose from among a range of options when faced with a threat to security:

> so that he can apply force adequate to accomplish the objectives at hand without causing any unnecessary damage or loss of life, and while holding to a minimum the risk of escalation to a more destructive level of conflict.

One way of achieving this was to acknowledge and work within the firebreak that divides nonnuclear from nuclear war, "a recognizable qualitative distinction that both combatants can recognize and agree upon, if they want to agree upon one." It would be prudent to agree, for no other obvious firebreak could be identified "all the way up the destructive spectrum to large scale thermonuclear war." Enthoven was aware of the argument for attempting to convince adversaries that the United States might use nuclear weapons even where this would be patently irrational, but he considered it foolish to base strategy on the pretense that the United States did not understand the destructiveness of modern war; not was this a viable course for a democracy.[39] This sort of analysis, and the language in which it was couched, was not revised during the McNamara period at the Pentagon. At least when thinking about a major confrontation with the Soviet Union, escalation remained a peril to be avoided.[40]

THE CUBAN MISSILE CRISIS

This sort of concern was very much in evidence during the Cuban missile crisis of October 1962. The popular commentary of the time, supported by some statements from the administration, implied that if the crisis could not be managed, then disaster might result. The president himself was extremely sensitive to the risk of a war getting out of hand, despite the beliefs of the participants that they knew what they were doing and had everything under control. According to one of his biographers, a course on the origins of World

War I at Harvard and a reading of Barbara Tuchman's *Guns of August* had left him aware of how nations could get dragged into war against their better judgment. "A favorite Kennedy word was *'miscalculation.'* "[41] Robert Kennedy's memoir of the crisis stresses the degree to which the risks of escalation impressed themselves on the participants. He describes a meeting with his brother and two advisers in the middle of the crisis:

> Neither side wanted war over Cuba, we agreed, but it was possible that either side could take a step that—for reasons of "security" or "pride" or "face"—would require a response by the other side, which in turn, for the same reasons of security, pride, or face, would bring about a counter-response and eventually an escalation into armed conflict. . . . We were not going to misjudge or miscalculate, or challenge the other side needlessly, or precipitously push our adversaries into a course of action that was not intended or anticipated.[42]

It was recognized by those debating United States options during the crisis that there was a risk of a Soviet response: a blockade would bring renewed pressure on Berlin; an air strike against the missile bases in Cuba could be countered by a similar strike against the Jupiter bases in Turkey. Some, including those advocating an air strike, assumed escalation was inevitable.[43] Others, including many in the military, felt that the risk of escalation was exaggerated and that the United States had more freedom of maneuver than commonly assumed, including the ability to remove Castro.[44] Marc Trachenberg has commented, on reviewing the transcripts of the crisis, that "What is striking here is the sense that peace was hanging on a thread, that it did not take much to set off a nuclear holocaust." McNamara in particular was "very sensitive to the danger of things spinning out of control."[45] The two sides certainly drew attention to the risk of letting the crisis get out of hand in their statements at the time. What is perhaps more impressive is the extent to which these concerns come through in their private communications. This may have been recourse to the threat of escalation as a source of leverage, but nonetheless reflected real concerns.[46] Trachenberg notes a genuine fearfulness over escalation and questions whether Kennedy saw himself as being engaged in a "competition of resolve."[47]

It was McNamara who came up with the idea of a blockade (later officially described as a "quarantine") as an intermediate step between a military and a political course. The president saw the need for a means to maximize the political support he expected after the existence of the missiles had been revealed: "The fact that we indicated our desire to restrain, this really would put the burden on the

Soviet." McNamara did not develop the idea of the blockade with great confidence: "Now this alternative doesn't seem to be a very acceptable one, but wait until you work on the others."[48] There were grave doubts in the United States as to whether the blockade would work, but more direct military action could be delayed until the missiles on Cuba were about to become operational. It did not preclude something more robust at a later date, while for the moment it shored up the political position of the United States.[49] Policy was, therefore, shaped by a combination of fear of escalation and the availability of a limited action that could reinforce the clear political message sent to the Soviet Union.

A variety of further steps were hinted at. In his speech of October 22, for example, the president described the quarantine of Cuba as just the beginning and directed the United States armed forces to "prepare for all eventualities" and the people for a crisis that could take "months" to resolve.

However, despite this and the later image of a preparedness to apply steadily increasing pressure through a series of graduated steps, there was no consensus in the Executive Committee that this was the obvious way to proceed should the blockade fail. It was assumed that the next step would be the previously discarded alternative of an air strike against the bases, or even a full-scale invasion. The crisis moved to a head on October 26 and 27. The United States enforced the quarantine and harassed Soviet submarines in the Caribbean: the Soviet Union rushed to complete the missile bases and shot down an American U–2 almost as soon as the SAM bases were completed on the island. The balance in the internal United States debate shifted toward the more drastic options. At this point, even McNamara considered an invasion "almost inevitable."[50] The president was still hoping to move in a graduated manner. He had resisted calls to strike the SAM bases and saw the next step as reinforcing the blockade by means of an expanded embargo list. This did not mean he was prepared for a contest of will and resolve. He sought to end the crisis through coupling a compromise solution (consideration of removal of the Jupiter missiles based in Turkey) with an ultimatum, which drew on the prospect of something other than a graduated response. When, on the evening of October 27, Robert Kennedy passed the president's letter to Soviet Ambassador Dobrynin, he stated:

> He should understand that if they did not remove these bases, we should remove them. . . . Time was running out. We only had a few more hours. We needed an answer immediately from the Soviet Union.[51]

Kennedy got the answer he wanted, and the crisis was resolved. The successful resolution of the crisis meant that there was an inevitable search for lessons that might be applied in any future crises. The president was one of those who cautioned against the temptation to generalize too much from this experience.[52] For one thing, Cuba was almost an ideal type: there were only two main players, though there were others in secondary roles; the issues at stake were relatively clear-cut, so there could be little doubt about the objective of military pressure; the military situation was also reasonably straightforward; it started at a definite time and did not creep up on the policy makers; and clearly had nuclear dimension, so there was little difficulty in imagining how the crisis might have escalated into something much worse.

Perhaps more seriously, even in these highly favorable conditions, the image of crisis management was somewhat at variance with the actual practice. For the purposes of this article, the most important image was of management through movement up a series of graduated steps in order to convince an adversary to back down. In practice, the graduated approach was only followed for a while and was coming under severe strain as the crisis closed. The final settlement came out of a sense of an imminent violent clash. The theorists had suggested that this fear of escalation would indeed produce a result, but they had not necessarily expected this sense to arrive so early. Moreover, the Cuban experience reminded of the importance of offering positive as well as negative inducements in achieving a settlement. The true lessons of Cuba for the strategy of coercive diplomacy, to which it gave such credence, were ambiguous.[53]

THE LESSONS OF CUBA

With the notable exception of the president, many of those agonizing over the appropriate course of action in the Executive Committee, which managed the crisis on the United States side, were the same people who began a similar if more prolonged quest for measures to arrest the deteriorating situation in Vietnam some eighteen months later. Before considering the extent to which their actions on Vietnam were influenced by the Cuban experience, it is important to note the effect on the strategic studies community.

For the civilian strategists, the lesson was not that the world had narrowly averted Armageddon, but that an extremely limited show of force had allowed the crisis to be satisfactorily managed. This point had been obscured by the overdrawn language in which the outcome of the crisis and the performance of the administration had been discussed. The management of this crisis appeared to validate the theories of the new strategists. Schelling enthused:

> The Cuba crisis raised the prospect of a very different species of "limited war." More than just raising the prospect, probably it should be construed as an actual instance of the new species, one in which no shots were actually fired. This new species is the competition in risk-taking, a military–diplomatic maneuver with or without a military engagement but with the outcome determined by manipulation of risk than by an actual contest of force.[54]

In an important paper published in early 1965, Albert and Roberta Wohlstetter took the opportunity to criticize the

> cliche and over-simple view that seems to have a special appeal in crisis . . . that the threat of force or the use of a low level of violence, including even a partial blockade, leads naturally to a higher level of violence.

For the Wohlstetters, the "cliché and over-simple view" was encapsulated by the idea of escalation, conventionally defined as "the spread and intensification of violence," which was assumed to tend toward all-out nuclear war. In questioning the automaticity of this process, the Wohlstetters noted not only the success of President Kennedy in reversing the direction of escalation in this crisis, but also how inaction—allowing the Soviet missiles to be established—could have led to much more dangerous consequences, and how the risks were, in practice, contained by the deliberate and tight control exercised by Kennedy and apparently by Khrushchev. The fact that the initial moves were backed by a "mounting sequence of threats short of nuclear war" backed by impressive conventional forces meant that there was no need for recourse to "desperate threats." In a critical passage, the authors took on the escalation metaphor:

> A process of escalation is usually thought of simply as an increase in violence growing out of a limited conflict in which an adversary may act to stave off his loss or an opponent's prospective success. The aspect of "escalators" that inspired its use in this connection, we suspect, is the fact that moving stairways carry a passenger on automatically without any effort of his will. However, as we have suggested there are down-escalators as well as up-escalators, and there are landings between escalators where one can decide to get off or get on, to go up or down, or to stay there; or to take the stairs. Just where automaticity or irreversibility takes over is an uncertain but vital matter, and that is one of the reasons a decision maker may want to take a breath at a landing to consider next steps.

In some ways, the Wohlstetters preferred a staircase to an escalator metaphor. In the case of Cuba:

> On the staircase of ascending steps in the use of force there would have been many landings, many decision points at which either side could choose between climbing higher or moving down.

The Wohlstetters were thus not arguing that there was no risk of a crisis getting out of hand; only that the risk arrived later than had been commonly assumed and that prior to that point there was a rich set of options.[55]

Bernard Brodie's approach, from a quite different angle, was also to counter pessimism on possibilities for controlling moves in war, though he parted company with most of his colleagues in believing that this was possible even with tactical nuclear weapons. His initial reaction to Cuba was to see in it confirmation of the possibilities of controlling escalation, when the consequences of letting matters get out of hand had become so apparent.

> We rarely have to threaten general war. We threaten instead the rest in a series of moves that seems to tend in that direction. The opponent has the choice of making the situation more dangerous, or less so.[56]

Brodie's main concern was to take on the administration's policy of developing nonnuclear options for NATO and of drawing attention to the firebreak that separated nonnuclear from nuclear war. He considered his policy misguided for a variety of reasons: (1) by stressing the nuclear firebreak it underestimated the significance of the most important strategic move—the original aggression that would cause the war—while overestimating the likelihood of such aggression; (2) it failed to capitalize on the clear Soviet doctrinal disregard of the firebreak and its keen awareness of the likely consequences of any war; (3) and it failed to appreciate the intensity of a nonnuclear war at the level allowed should there be strict adherence to the firebreak.

He had come to consider the threat of limited nuclear use to have far more deterrent value than he, among others, had assumed it had in the 1950s. This was partly because of technological developments, which he overstated, and partly because the strategic balance was turning out to be sufficiently *indelicate* to dampen down escalatory pressures. In addition, Brodie was bolstered by an understanding of Soviet strategy as one of testing and probing the West, yet quick to back down on meeting firm resistance.

In developing this position, Brodie felt obliged to address the issue of escalation. Steiner records an early use of the term in a letter in March 1960, when it was defined orthodoxly as the "unintended spread of nuclear war."[57] He did not analyze the concept fully until a September 1962 RAND Working Paper. In this, he noted that the automaticity suggested by the conflict need not be inevitable:

> "escalation" usually refers to a change from limited to general war. However the term itself connotes the idea that progress from one condition to the other can be by stages, which may, however, be traversed rapidly.[58]

In 1966, he published *Escalation and the Nuclear Option*, his most substantial discussion of the topic. This largely took the form of a critique of the administration's desire to reduce the risk of escalation to general nuclear war by means of stressing the firebreak. It should be noted that Brodie's definition of *escalation* had not altered—"the erosion or collapse of limitations"—although at one point he distinguished between "deliberate" and "uncontrolled" escalation. He observed that with the easing of the fear of surprise strategic attack, this process was now the most likely cause of a general war. It had become feared that escalation would automatically follow any outbreak of war. The arrival of strategic stability reduced the general risk of escalation, even as a result of tactical nuclear use.[59]

Leaving aside the validity of this judgment, Brodie's understanding of the dynamics of escalation are worth considering. The presumption of uncontrollable escalation, he suggested, was based on:

> (a) a well nigh limitless concern with saving face, regardless of risk, (b) a great deal of ground-in automaticity of response and counter-response resulting in a swiftly accelerating ascent in the scale of violence.

Such tendencies existed, but not in a vacuum. It was the "mechanical" view of warfare to which he objected, the assumption that responses were so rigid that the outcome was predetermined. Those responsible would not necessarily succumb to the "bundle of psychological factors summed up by (a) loss of face and (b) tendencies to yield to feelings of hatred and rage" that were said to explain escalation. To assess the genuine risks of escalation, it was necessary to predict "the behavior of their leaders (as well as our own) under various kinds of crisis situation." So "the control of escalation is an exercise in deterrence," and as Brodie was extremely confident about deterrence, he was confident that escalation could be controlled. Brodie, therefore, did not challenge the tragic concept of escalation, only the presumption of inevitability once the first nuclear shots had been fired, or indeed any shots.[60]

THE CONTROL OF ESCALATION

So, whereas the degree of concern over escalation was challenged, the concept itself was not contested. It was recognized to refer to a process by which a limited conflict might turn into something more violent and dangerous. At issue was not whether this process might materialize in war, but under what circumstances. The title of Brodie's book suggested that escalation was a problem to be considered in contemplating the nuclear option. The Wohlstetters' "Adelphi Paper" challenged the escalation metaphor head-on; it did not appropriate the term in favor of what was soon to be described as "controlled escalation."

Even in the first collection of essays that took seriously the possibility of controlled nuclear use for bargaining purposes, edited by Klaus Knorr and Thornton Read, escalation was generally described as the antithesis of control. Herbert Benington, for example, listed escalation as one type of instability—along with accidents, unauthorized behavior, confusion, and miscalculation. Both the editors used it in this way. With Knorr, the adjectives *rapid, unlimited,* and *violent* were added at different times to convey the extent to which this is something to be avoided. Read referred to the "built-in" escalation mechanism" of tactical nuclear weapons.[61]

There were, however, two dissenters to this tragic view: Herman Kahn and Morton Kaplan. Kaplan was one of the first writers to talk of escalation in terms of deliberate and positive strategic moves. He described the escalation process as a graduated series of acts, rather than as a process running out of control. An act of escalation had to be understood in terms of its effects on the perceptions of the other side:

> As each considers escalation, either to prevent defeat or to oppose the opponent's escalation, it must consider the consequences of the case in point for other cases and ways in which its actions will shape its opponent's expectations.

If escalation was going to be exploited for strategic advantage, then there had to be a degree of mutual understanding as to its meaning. It would be necessary for the "USSR to accept the American criteria for the circumstances in which escalation may be employed." (The use here indicates that escalation was a deliberate matter and not an involuntary process.) Kaplan also introduced the notion of what later came to be known as *escalation dominance*: "We may be able to escalate to a level of action that is comparatively disadvantageous to the Russians."[62]

It was Kahn's name that eventually became most associated with such ideas. In this volume, his discussion of escalation largely followed the general use of the term as an uncontrollable process concluding with all-out war. It was something to be avoided, although at an early stage on the "escalation ladder" (a term introduced in this essay) it might be possible to use the fear of escalation.

> One can never be certain when a difference of opinion will not turn into a dispute, a dispute into a minor crisis, a minor crisis into a more serious crisis, and so on ending in disaster. In fact the threat of such escalation and its ultimate outcome—mutual homicide—may sometimes be used deliberately.

This was very similar to Schelling's discussion of the same idea in *Strategy of Conflict* and, indeed, in his contribution to the Knorr and Read volume ("raising the risk of general war is part of a strategy of conducting limited war"). By and large, escalation still appeared as something to be avoided, but Kahn is clearly becoming intrigued by the possibilities for control.[63]

In *Thinking the Unthinkable*, published later the same year, Kahn moved the concept along further. The first reference to the "process generally called 'escalation'" provides an orthodox definition: "A limited move may appear safe, but set into motion a disastrous sequence of decisions and actions."[64] However, in one chapter, "International Bargaining and The Game of Chicken," he develops the idea of an "Escalation Ladder" (always in capitals!). This was one of the classic oxymorons of strategic studies. It brought together in one metaphor the two alternative and very distinctive means of traveling in a similar direction, reflecting fully the tension between automaticity and control, inadvertence and deliberation. As the title of the chapter indicates, Kahn was contrasting his understanding of the possibility of applying force as an exercise in bargaining with the fear that a superpower contest could degenerate into the equivalent of the juvenile game of chicken. Kahn wrote of an escalation ladder because he wished to emphasize that this was a stepped process.

> Each side may take certain positive steps either to bring the other to the bargaining table or to apply pressure during the negotiations. Sometimes these pressures tend to decrease with time or with a

> temporary solution to the problem at hand. At other times there is a tendency for each side to counter the other pressure with a somewhat stronger one of its own. This increasing pressure step by step is called "escalation."[65]

Thus it was not taking a step that constituted escalation, but the process that was set in motion when the other side responded with a step of its own. Escalation referred, then, to a process made up of moves and countermoves rather than a single act, and this process could develop rapidly and violently or more gradually. It depended on the interaction of moves taken by two opposing sides. A decrease in the level of violence Kahn later described as "de-escalation"—presumably the consequence of a particular step upward having served its purpose in influencing the adversary. But once Kahn began to disaggregate this process into a series of deliberate, interacting moves, the sense of automaticity was lost. Each step then became an act of escalation. It is not a large leap from escalation-as-process to escalation-as-strategic-move.

Nevertheless, Kahn was still operating with the idea of escalation as a set of pressures tending toward all-out nuclear war. Nor was he claiming to be describing all the steps through which all conflicts might potentially pass en route to all-out war. He opened the discussion with a series of qualifications, which are also made at great length in his later book, explaining why the ladder should not be taken *too* seriously. It was intended to "prod the imagination, not confine it."[66] The ladder had sixteen rungs. Only the last five involved actual combat. With initial nuclear use, he stressed the importance of signaling "resolve and committal" rather than "to destroy the other side's military forces or to handicap his operations." There was hardly any discussion of escalation *within* nonnuclear war.[67]

By 1965, the ladder had acquired forty-four steps and justified a volume all its own—*On Escalation*. Now escalation had been turned into a deliberate act. The sort of uncontrolled process that had once been described as escalation was now "eruption." Although a whole chapter was devoted to drawing attention to problems with the escalation ladder (now no longer in capitals), the comprehensiveness of the list supported the view that it represented a "spectrum of violence."

In a useful appendix, "Relevant Concepts and Language for the Discussion of Escalation," was the following discussion:

> The term "escalation" focuses attention on the fact that each side can increase the stakes, and that bargaining is almost always crucial: that deterrence continues to function during and after an attack as

> well as preattack, and is likely to be as important as denial in preserving a country.

Later in this appendix, he defined the influential notion of "escalation dominance." The side with such dominance has a marked advantage in a given region of the ladder:

> It depends on the net effect of the competing capabilities on the rung being occupied, the estimate by each side of what would happen if the confrontation moved to other rungs, and the means each side has to shift the confrontation to these other rungs. One variable affecting escalation dominance is each side's relative fear of eruption. That side which has least to lose by eruption, or fears eruption the least, will automatically have an element of escalation dominance.[68]

To most people, the striking feature of the book was the possibility of anyone coming up with almost thirty distinct steps following the first use of nuclear weapons at rung fifteen.[69] Analytically more useful was the identification of six firebreaks or "thresholds" at which very sharp changes in the character of the escalation took place. These were: (1) don't rock the boat; (2) nuclear war is unthinkable; (3) no nuclear use; (4) central sanctuary; (5) central war; and (6) city targeting.[70] The book was taken up with an examination of the various tactics that might be available at each of these rungs. It was not always clear what relevance action at one level had for later escalation. Because Kahn was not attempting to predict the course of any given conflict, there was no necessary link between one level and another. His basic objective was to demonstrate the potential for cool, considered action at all times, with a political as much as a military purpose in mind. Nevertheless, the impression created was of a gradual movement along a continuous "spectrum of violence," with each new escalation undertaken because the previous one had proved insufficient or had been matched by the other side.

Kahn thus challenged the original concept of escalation by offering a new meaning. An alternative approach to handling a similar set of ideas was provided by Morton Halperin's 1963 survey, *Limited War in the Nuclear Age*. Halperin sought some conceptual clarification by discarding rather than refining the term escalation, and replacing it with two alternatives. Both of these were already in the literature. They allowed for a distinction between two separate processes: *expansion*, a gradual increase in the level of military force employed in a conflict, and *explosion*, the sudden transformation of a nuclear war into a central war by the unleashing of strategic nuclear forces. Of these terms, *explosion* was the closest to the prior use of *escalation* (we noted earlier that the two terms were used interchangeably for a time). According to Halperin, expansion was likely to occur before an explosion. This was the period when matters were generally under control and it was still possible to exploit the fear of them getting out of control.

Halperin linked explosion very closely with a strategic first strike. He distinguished between two types of explosion; *inadvertent* and *deliberate*. The former related to the aggravation of all the pressures to strike first during the course of a conflict (for example, along the lines of Schelling's "reciprocal fear of surprise attack"); the latter to a clear decision by the losing side on a battlefield to initiate a strategic strike rather than accept defeat. Therefore, the risk of explosion was likely to be as much a function of the stability of the strategic balance as of inherent tendencies in warfare. Halperin's belief that the strategic balance was becoming more stable led to the judgment that the risks of explosion were declining and were likely to be perceived as declining by the major powers, in which case, one of the key restraints on expansion would be relaxed. Thus:

> It is conceivable that a war the size of World War II could be fought without it expanding or exploding into central war. Certainly a major restraint against wars of the size of Korea will be eliminated when the pressure against expansion, stemming from the fear of an explosion has been reduced. This very lack of pressure against the expansion of a local war might, unless other pressures intervened, lead to a war which very slowly and gradually expanded to the point where it became a central war. One might imagine, for example, a war on the European continent becoming larger and larger until it led to strategic attacks on the Soviet homeland which were followed by attacks on the United States. In such a war, however, the major powers might still observe a number of important limits including not attacking cities.

Halperin's analysis depended on a stable strategic balance serving as a powerful counter to any tendencies to explosion. In no circumstances could a strategic first strike improve matters for either belligerent; it would almost certainly make them worse. This was very similar to the position adopted by Bernard Brodie: that the pressures toward escalation would be countered by the knowledge that there was nothing to be gained and much to be lost by moving toward extensive strategic nuclear use. This suggested a process still largely under control, a view underscored by Halperin's presumption that it would still be a matter of expansion when the conflict moved to something on the order of World War II (which only because of the new association of "strategic nuclear" with "central" war could be deemed limited).

Yet Halperin also recognized that the danger of explosion could arise at any time in a war undergoing expansion. And he did not assume that wars naturally stay within limits: "Unless there is an explosion, the process of expansion will continue until both sides decide that it is not in their interest to expand the war." It was the political context—the local factors that prompted the conflict in the first place and the wider relationship, including alliance obligations—in which the major powers were conducting a limited war that determined the circumstances of war termination. Thus, expansion could not be seen simply as a matter of the moves made by the two sides to improve their tactical military situation, and the resulting countermoves. It was also a function of the bargaining between the two sides. This bargaining concerned both the character of the limitations that might be observed and the questions of substance at the heart of the conflict. The danger of explosion in this process was reduced so long as both sides had some appreciation of the limits within which they were operating, or, more to the point, recognized certain limits that, if transgressed, heightened the risk of explosion. So in principle, the more the potential for explosion was controlled, the greater the potential for expansion because the risks had been reduced. This confident analysis depended on the policy makers taking a similar view of the risk of explosion.[71]

By 1967, Halperin had acknowledged that the desire of the major powers to avoid a general war had served as a source of caution. Although he still assumed that a stable strategic balance should reduce the risk of expansion, he noted that

> political leaders . . . have tended to overestimate the probability that nuclear war would develop from any particular local crisis and, for that reason, have been less willing to commit military forces than they might otherwise be.

In the absence of a clear image of how a general nuclear war might arise out of a local war, political leaders impressed by the risk that

> growing tension and increased use of military force could somehow get out of hand and lead to a nuclear conflict which nobody wanted. They have realized that when decisions have to be made more quickly, when troops have to be authorized to make some decisions locally, and when tensions and political conflicts heighten, events may no longer be under the complete control of political leaders and may generate pressures leading to a nuclear war.[72]

In *On Escalation*, Kahn picked up Halperin's terminology, noting however that he preferred to talk about "eruption" instead of Halperin's "explosion." Kahn described Halperin's "expansion" as "escalation." Here, then, we have the twist. Instead of a synonym for *eruption* or *explosion*, *escalation* was now used to describe a process that was much more controlled and deliberate.[73]

THE NEW IDIOM OF MILITARY FORCE

There was always a tension between the new strategists' belief in the capacity of national leaders, properly advised, to exploit efficiently and effectively America's power against a powerful adversary, and the tolerance of a concept that stressed the uncertainty and uncontrollable aspects of war. Escalation initially appeared as a challenge to the strategists, a description of exactly the sort of wild momentum that had been so difficult to tame in the past, and yet had to be tamed now if strategy was to have any meaning. Nuclear weapons were the new element that made the prospect of escalation so frightening, yet they were also seen as the best hope of draining the escalatory process of its inner drive. When the final destination was so patently awful, then it would be best not to travel. The dilemma was that a degree of traveling was necessary to convince the adversary that the United States was serious. Any show of resolve had become a risky business, so the challenge for the strategists was to devise controlled means of demonstrating resolve. The Cuban missile crisis confirmed to the strategists that this challenge could be met.

The application of military force in these new circumstances required a new idiom, so that the linkage with the political processes of bargaining and negotiation would be more explicit. The strategists disagreed among themselves on the extent to which nuclear use could conform to this pattern, or would simply degenerate into massive destructive exchanges. They tended at least to agree that nonnuclear actions would gain much of their impact by drawing on this prospect. The presumption throughout was that at issue was the management of a series of contests between two nuclear powers.

The paucity of serious thought on what might be happening in the conventional stages is striking. Action in a crisis was significant in terms of what it conveyed about a willingness to run nuclear rather than conventional risks. A readiness to run conventional risks was taken for granted and lacked the political impact of nuclear risk taking. The traditional concern of the military with engaging the enemy on the battlefield, and deciding matters through force of arms, with the accompanying requirements for secure supply lines and reserves, equipment, and training, had all but become irrelevant as a strategic problem. Battle in the traditional sense could achieve little; a single nuclear strike could mock its conclusion.

Schelling observed in the mid-1960s that:

> Most of us, in discussing limited war during the past ten years, have had in mind a war in which *both* sides were somewhat deterred during war itself by unused force and violence on the other side. . . . We have mainly been talking about wars that involved some *continued* mutual deterrence, some implicit or explicit understanding about the non-commitment of additional force or non-enlargement to other territories or targets.

However, he added, even if one faced "no possibility of enemy escalation," there was a reason for withholding forces: "It is simply that the object is to make the enemy behave."[74]

VIETNAM

The implication was that the concepts of coercive diplomacy could be applied in circumstances quite different from those of a superpower confrontation. Vietnam offered the first major test after Cuba to discover whether or not this might be true. The escalation took place beneath the nuclear threshold and against an adversary other than the Soviet Union whose responses to the various American stimuli were harder to predict. Certainly, many of those responsible were influenced by the Cuban experience and may have believed that they were applying the lessons of this experience.[75]

If this was the case, then the Vietnam experience soon warned that it would be unwise to generalize too much from the success with the Cuban missiles. Robert Osgood, the first postwar American theorist of limited war, described the strategy of "controlled escalation, as applied by means of selective bombing in Vietnam" as being close to a "clear-cut failure."[76] Osgood is one of the theorists held responsible by Stephen Rosen for creating an over confidence in the possibilities of strategic communication, and a "relative disregard for the military and domestic political problems of waging war."[77]

It has been argued by some associated with the air campaign that the theorists of graduated response completely misunderstood the nature of air power, which achieves its results through decisive blows rather than steady pressure.[78] But was it simply the case that the Vietnam failure was exacerbated through the application of an inappropriate strategy, or was it that the strategy was not even applied, and possibly could never have been under the prevailing conditions? Were the ideas of graduated response, that later came to be described as controlled escalation, influencing the policy makers? More to the point, were they influencing policy?

While the United States had been steadily building up its involvement in Vietnam during the early 1960s, the civilian strategists had taken little interest in its course. Their theories only became relevant during the intensive discussions throughout 1964 on what to do about the deteriorating situation in South Vietnam. Up to this point, if Vietnam had been a testing ground for new strategic concepts, it had been those of counterinsurgency. Now the question arose as to whether counterinsurgency by itself could ever be sufficient, and whether something would need to be done against the North.

These deliberations began in earnest at the start of 1964, stimulated by military concern over the communist advances in the South. In the many internal studies, the ideas of graduated response and the experience of Cuba were clearly influential.[79] The debate itself was inconclusive, with the pressure for a more substantial action countered by anxiety over the risks attached to such action and the political requirements of the 1964 presidential election. In August 1964, there were air raids against the North undertaken as reprisals for attacks by North Vietnamese patrol boats against the destroyer USS *Maddox* in the Tonkin Gulf, which established both the precedent and the congressional support for later raids. However, there were no comparable reprisals in November 1964, after an attack on the United States air base at Bien Hoa, and the next month, after an attack on the American officers' billet in Saigon. Later—in February 1965—the Communists themselves further raised the stakes, including by a raid on the American billet at Pleiku. After an intensive internal debate, the president eventually decided to authorize air strikes against the North as well as the limited introduction of two Marine battalions. Operation "Rolling Thunder" began on March 2. It was something of a compromise response and failed to induce any changes in the attitude of the North Vietnamese. The main result was to raise the visible level of the American commitment, which, it was soon discovered, required a substantial military effort.

In accounts of United States decision making in 1964 and early 1965, attention is often drawn to the influence of theorists such as Schelling and Kahn and the notion of "coercive diplomacy."[80] This is presented in terms of either the general intellectual context or the influence of specific individuals. John McNaughton, Robert McNamara's main adviser on Vietnam, is sometimes singled out as the carrier of these misguided notions. Fred Kaplan traces McNaughton's approach back to the ideas of his friend Thomas Schelling.[81]

The academic strategists certainly recognized some aspects of the theories in the events going on in Vietnam. Herman Kahn, writing after this incident but before the bombing campaign of 1965, suggested that as a result of this success "one can almost predict that the U.S. will be more willing to escalate in North Vietnam."[82]

In *Arms and Influence*, Schelling also found the Tonkin Gulf reprisals "articulate" and "appropriate," but he expressed himself unsure about the "coercive aspects of the bombing of North Vietnam." Although he acknowledged the bombing as part of a class of actions that he had helped identify, this was a quite different situation, namely, the character of the enemy (including his lack of capacity to inflict direct hurt on the U.S.), the position on the ground in Vietnam, and the consequent importance of the adversary's military assets as much as his vulnerable population as a target for United States coercion. Schelling noted the criticism of the diplomacy accompanying the bombing and hoped that there had been private communications to Hanoi of a less ambiguous nature. On the other hand, he felt that the signal of "unintended enlargement" could have warned off the Chinese.[83] He clearly did not wish the Vietnam buildup to be accepted as a clear application of his theory.

Within government, much of the language in which these issues were discussed testifies to the influence of the new strategists, but the underlying themes were not that novel. The past experience of the participants encouraged them to think as they did, just as much as the newfangled ideas emerging out of RAND and the Hudson Institute. The fact that Walt Rostow spent much of World War II picking targets for the Army Air Force, and George Ball assessed the consequences as a member of the United States Strategic Bombing Survey, probably explains much about the different views over the efficacy of air power as an instrument of changing the will of an enemy. Those who had been through the Cuban missile crisis remembered that a limited, measured action had succeeded and that the judgment of those military men prescribing stiffer measures had been shown to be faulty—possibly dangerously so. Those thinking back to Korea, the starting point for all United States thinking on limited war, were reminded of how real the risks of drawing in the Chinese might be—despite military disclaimers—and how unfortunate the possible consequences. In this sense, the theorizing was influential not so much in designing a theory of graduated response, but in encapsulating the relevant experience of the previous decade and a half and so providing a good reason to keep the war limited. One of those people involved has commented:

> My sense of the Vietnam escalation debate is that it was conducted almost wholly in its own specific terms with little reference to, or self-conscious awareness of, outside thinking.[84]

It is of note for the purposes of this study that references to escalation, in the sense that it came to be used after this period, are comparatively rare in the relevant documents. *Escalation* was not used to describe a strategy of controlled pressure on an enemy, but rather a situation to be avoided in which the level of military action was being raised too high. The connotation was still essentially of a loss of control. At the time of the critical decisions of 1965, we find McGeorge Bundy arguing for "A Policy of Sustained Reprisal" and Maxwell Taylor for "Graduated Reprisals," contrasted with Admiral Sharp's preference for "Graduated Pressure" and John McNaughton's "Program of Progressive Military Pressure." In the latter's memo, the question of escalation is only raised with consideration of the "Red flash points" which implied "a substantial risk of escalation with China or the Soviet Union."[85]

As an influence on actual policy, and not on individual policy makers, the theories of coercive diplomacy emerge as at most one of the many factors at work at the time. The other factors were often of a nature to preclude the application of military power according to the dictates of these theories. For a start, there was no consensus as to what a campaign of pressure against Hanoi was designed to achieve. Some argued in terms of forcing the North Vietnamese to withdraw support for the Viet Cong in the South; others were largely concerned with interfering with the infiltration of men and supplies from the North to the South. A third objective was to force Hanoi to negotiate on American terms, which would indeed have been the logic of a true strategy of coercive diplomacy. However, the strength of the communist position in the South meant that there was no basis for a settlement in terms of the status quo, and anything more favorable would require success in the counterinsurgency effort. In denying extra supplies to the Viet Cong, air raids against the infiltration routes might have helped, but there were doubts that the Viet Cong were that dependent on these supplies or that, to the extent that they were dependent, the raids would effectively stem the flow.

This demonstrated the difficulty in simply viewing counterinsurgency as the first in a series of graduated steps on the ladder. It might have been so in terms of the extent of American commitment, but it was crucial to everything else that followed. A failure to stabilize the situation in the South could not be remedied simply by upping the ante and fighting a different sort of war. So long as the antigovernment forces were strong in the South, the United States bargaining position in the search for a settlement would be weak, and the required degree of coercion of the North would be correspondingly greater. The second problem was that when contemplating a greater American involvement, the United States did not possess the inner confidence stemming from a willingness to go the distance with the other side. There were serious concerns over the prospect of drawing China in on the North Vietnamese side, as had happened in Korea, and a reluctance to contemplate sending in American ground forces. The United States could not contemplate escalation with equanimity. The American commitment was therefore qualified.[86]

The dilemma in 1964 was that some way had to be found to close the gap between the stated American commitment to the defeat of the communist challenge in the South and the continuing success of that challenge. the Johnson administration was unwilling to wind down the commitment. The measures that were eventually chosen were, if anything, more affirmations of a retained commitment, directed not so much at Hanoi but rather at the morale of the South Vietnamese government and its armed forces, as well as the opinions of other allies and friends. Because United States military power had not been seriously tested, there may have been a feeling that the United States had only to flex its muscles and the enemy would crumble at the prospect of facing American power. But there was no clear strategy to enforce this, and when United States muscles were flexed, the North Vietnamese responded by increasing their own efforts.

It is important to recognize what was being expected of Hanoi. At a time when the United States was increasing its support for the government in Saigon, the North Vietnamese were being required to desist from supporting the Viet Cong, and so accept a rolling back of the gains that they had made in recent years. Yet this requirement was implied rather than clearly stated. As William Simon has observed:

> At no time did the United States government explicitly threaten the DRV [Democratic Republic of Vietnam] with increasing punishment or suggest eventual damage to its industrial capability if it did not comply. Neither did it attempt to convey any sense of urgency by indicating a period of time during which compliance would be expected.[87]

The problems were illustrated by the Tonkin Gulf episode. The raids were agreed, at the time, to have been effective in domestic political terms, and even in making a more general point about United States resolve and commitment. In practice, they appeared to stimulate an expansion of the North's effort rather than a prudent restraint. The manner in which the United States justified these raids —as a reprisal—warned against further provocations of this type rather than the consequences of the North's failure to reduce the general level of support for the insurgency in the South. This brings out a tension in the application of the theory. It was discussed as a means of forcing the pace of a conflict, yet the issue was as likely to arise in terms of the need to respond to enemy moves, or at least was as likely to be presented in terms of a proportionate response.

One notable feature of the debate in Washington during this period is that many of these problems were recognized by the participants. A good example of this is a study, "Alternatives for Imposition of Measured Pressures against North Vietnam," prepared under the direction of a member of the State Department's Policy Planning Council by an interdepartmental committee, in response to a request by the Council's Chairman, Walt Rostow. Rostow, who was a great believer in strategic air power and taking the war to what was supposed to be the source of the insurgency, had been arguing since 1961 in favor of bombing the North. The study examined this issue as a practical policy option from all angles and in all variations. It was highly sceptical.

The primary objective of the proposed strategy of measured pressures was described as "a reduction in the level of Communist activity in Laos and South Vietnam to the point where our counterinsurgency strategy could succeed." The method would be to exploit:

> (a) Communist fears that military action might escalate to levels highly dangerous for them, (b) DRV concern lest the rudimentary but cherished industrialization in the North be wiped out, and (c) DRV apprehensions that increasing Chinese involvement would cost them the control they have struggled so long to gain over the territory and the people of North Vietnam.

At the same time, the strategy would seek to exclude Chinese or Soviet involvement and rejuvenate the counterinsurgency effort in the South.

The options available would be (1) to carry on with the current set of covert, nonattributable activity; (2) to move to more overt activities, such as a higher profile United States military presence in the area; and (3) to include direct United States attack on DRV territory, of which air attack was one of the most likely. Underlying each alternative was "the broad concept that military actions, whether combatant or not, would be taken for the political purpose of channeling the decisions of DRV leadership."

The study highlighted a series of problems with this strategy: it cast doubt on whether the primary objective could be achieved by this method, at least without more action on the ground by the United States and a greater degree of "dynamism" by the South Vietnamese forces; and it warned of the potential costs, including those of possible failure and condemnatory action in the United Nations. In these circumstances, negotiations that would be of value in putting the United States case would become extremely difficult. If Hanoi claimed to be complying with United States demands it would be extremely difficult to measure this in a convincing manner, but without this how could one justify a cessation of pressures by the United States? If, on the other hand, the pressure was not producing results, how could the United States withdraw without escalating further? One risk was of major Communist escalation. Although considered unlikely unless the survival of North Vietnam was threatened, it could "raise questions of introduction of US ground troops, possible war with Communist China and nuclear weapon use."[88]

The paper was highly classified and did not gain wide circulation. However, its influence can be seen in George Ball's famous dissent of October 1964, which drew on an understanding of the processes of escalation to make its point:

> It is the nature of escalation that each move passes the option to the other side, while at the same time the party which seems to be losing will be tempted to keep raising the ante. To the extent that the response to a move can be controlled, that move is probably ineffective. If the move is effective, it may not be possible to control—or accurately anticipate—the response.
>
> Once on the tiger's back we cannot be sure of picking the place to dismount.[89]

CONCLUSION

The Vietnam experience naturally led to a serious reassessment of the theories of military power that had become so fashionable early in the 1960s. It is of note that some of the more impressive works to come out of RAND toward the end of the 1960s and early 1970s were more skeptical and less self-confident, drawing on historical experience as much as abstract theory.[90] One example is Alexander

George's dissection of the concept of coercive diplomacy, which had begun as a briefing to the Air Force in the middle of 1965 on the sensitive matter of why the bombing campaign against the North had failed to produce the expected results. George drew attention to the conditions under which coercive diplomacy might operate, stressing the importance of a series of political factors such as, motivation, sense of urgency, domestic political support, clarity of objectives, and precision in defining an acceptable settlement, as well as the existence of viable military options and the adversary's fear of escalation.[91]

The problem was that the theory was more dependent than realized on the political and military context in which it was being applied. A conflict's limits, and threats to move beyond those limits, could not be guaranteed to emerge out of the inner logic of a conflict but needed to be spelled out in detail through some form of diplomatic communication. Moreover, the political signals transmitted by military actions were often geared to a domestic and wider international audience as well as the adversary.

It had been assumed that United States decision makers would have to hold back pressures for reckless action, but little account had been taken of the need to convince a sceptical public or allies that some actions were justified. It had been assumed that the United States would be operating from a firm base rather than being continually preoccupied with the weakness of the ally it was seeking to defend. Success in conflicts of this type had been shown to depend not only on establishing limits on the conflict or the ability to manipulate the risk of moving beyond the limits, but also on the ability to fight effectively within those limits. As Stephen Rosen has commented:

> The general problem of limited war is not only the *diplomatic* one of how to signal our resolve to our enemy, but the *military* one of how to adapt, quickly and successfully, to the peculiar and unfamiliar *battlefield* conditions in which our armed forces are fighting.[92]

What relevance did this experience have for strategic thinking with regard to a major United States–Soviet confrontation? It was somewhat easier—as a practical matter—to devastate Soviet cities than to help shore up an inefficient and unpopular government against a determined opposition backed by a highly motivated and mobilized neighbor. This suggested that it was much easier to apply theories of strategic bargaining to a set piece, East–West confrontation than to counterinsurgency. However, it was not good enough to argue that the two cases were wholly different.

Nuclear use, or at least the more substantial military signals, was presumed to involve a relatively simple set of variables, with the process moved along by a straightforward mechanism of action and reaction. Reprisal and retaliation were the key words. The two sides would be in many ways symmetrical in their capabilities. The practicalities of implementation were not at issue: after all this was the age when there were *no* limits on the crude exercise of military power. The sort of natural limits on warfare provided by geography, topography, demography, finance, and transport were becoming less relevant. At issue for the limited war theorists was whether appropriate limitations could be contrived for nuclear use.

The frightening prospect of a war without obvious limitations distracted attention from the dangers and difficulties likely to be faced in wars fought within clear limits. Because American power was unlimited, it was assumed that there should be no problem fighting within whatever limits had been applied.[93] Yet once limits had been established, then questions of comparative military advantage within those limits became critical. The fact of superior power beyond those limits became almost irrelevant. Any move beyond the limits had to take account of the countervailing power of the enemy. Because such moves would be dangerous for all concerned, what was believed to be at stake in the conflict would be as important as military capabilities.[94]

The Vietnam experience was certainly taken as a salutary lesson with regard to the pitfalls of a policy of deliberate escalation. It also had a lasting effect on the concept itself. Morton Halperin's discussion of the difference between *expansion* and *explosion* was noted earlier. Expansion was the more gradual and controlled of the two processes. The early use of *escalation* in the strategic studies literature (and the one current at the time of the 1964–65 debates on Vietnam) was equivalent to Halperin's *explosion*. Kahn's use of *escalation*, however, was much more similar to Halperin's expansion.[95]

In no small measure because of Kahn's efforts, *escalation* was very much part of the strategic vernacular when the first postmortems on the policy failure in Vietnam began. It was the natural term to be used in explaining both what the participants thought they were doing and where they went wrong. The policy debate on action against North Vietnam and the subsequent academic analysis of this debate brought about a different sort of shift in the understanding of escalation. The use of such words as *graduated* and *measured*, which were applied to reprisals or pressures, was noted earlier. Something very controlled and deliberate was suggested, proportionate to the enemy transgression, along the lines (though on a smaller scale) of Halperin's expansion. There were references to escalation, but as something much more substantial and risky. It did not refer so much to an independent process beyond deliberate control but rather to actions taken that heightened the risk of a loss of control. The connotation was, therefore, of a disproportionate response that could move the level of conflict to a completely new and dangerous plane.

The experience of Vietnam suggested that gradualism and escalation were not as distinct in practice as they might appear. Upping the ante, to coerce or simply to gain a military advantage, was always likely to be subject to dangerous uncertainties, and because of this it was always likely to be deemed disproportionate. So those proposing limited moves in the mid-1960s, designed to advance United States objectives without taking undue risks, described graduated and measured acts of pressure or reprisal. Kahn was prepared to call these moves escalation. Those looking back on the experience also called them escalation, but simply to warn of the extent to which these moves tended to be more substantial and dangerous than supposed by their proponents.

Three sorts of problems with escalation might be thought to have been illuminated by the Vietnam experience. The first, which we have already considered at some length, was the belief that the threat of escalation could be used to coerce an enemy. The second was the "seduction of gradualism," the belief that the extent of a military commitment could be controlled by moving forward only in short and discrete steps.[96] However, Vietnam did not represent an escalation process out of control, in the sense that the word had been used originally to describe the transformation of a limited into a general war. Each new step the United States took was deliberate and considered, and it eventually proved possible to de-escalate. The worst fears of drawing in the Chinese Communists or moving to nuclear use were not realized.

The third problem was the sheer difficulty of fighting within limits, both because of the real constraints on military operations posed by the limits and the prevailing sense of nervousness that the limits, proximate and distant, would nonetheless prove to be impossible to sustain. The theorists of escalation had concentrated on the ultimate level of violence, against which any violence at lower levels seemed tolerable. They had assumed that both sides would fight at these lower, tolerable levels so long as the risks of escalation could be controlled. This view had failed to take account of the particular difficulties, dangers, and frustrations to be faced at conflict at any level, and the extent to which the thought of what even the most modest encounter might lead to had dampened enthusiasm for conflict at any level. The risks of escalation now appeared to attach to *any* military move.

By the second half of the 1960s, *escalation* had lost its original, precise meaning as a description of the great risk involved in any use of nuclear weapons against another nuclear power. It had begun as a description of a process that governments might not be able to control and that would end in disaster. It was then turned into a device for exploring the possibilities for control, at a time when the United States, and its intellectual community, were most confident about their ability to manage the most intense, passionate, and dangerous of human activities. It concluded as a description of a deliberate move likely to intensify any conflict. Its negative connotations stemmed from both the dangers to be faced at the new level of conflict and the ever present fear that once one limitation on conflict had been eroded, further limitations would prove progressively difficult to sustain. When contemplating how one might ride a tiger, the most prudent course appears to be not to mount it at all.

NOTES

1. Richard Smoke, *War: Controlling Escalation* (Cambridge: Harvard University Press, 1977), p. 35.

2. Ibid., p. 19. Colin Gray suggests that escalation, along with "crisis management" was only briefly in the strategic "hall of fame" because, as a concept, it was "inherently subordinate and derivative, unpromising for development," and that although it became a household word, it did not benefit from "rigorous scholarly attention." Colin S. Gray, *Strategic Studies and Public Policy: The American Experience* (Lexington: The University Press of Kentucky, 1982), p. 108.

3. *New York Times*, March 29, 1961.

4. Fred Ikle, "When the Fighting has to Stop: The Arguments about Escalation," *World Politics*, July 1967. It is not clear whether Ikle was aware of any professional use of the term in the early 1950s. The fact that he still described the term as "scarcely fifteen years old" four years later suggests that he was not being especially precise. *Every War Must End* (New York: Columbia University Press, 1971), p. 39. Morton Halperin, in a book published the same year, noted how the term had " 'escalated' into everyday vocabulary from the more technical vocabulary of the strategic analyst." Morton Halperin, *Contemporary Military Strategy* (Boston: Little, Brown & Co. 1967), p. 22.

5. Morton Halperin, *Limited War in the Nuclear Age* (New York: John Wiley, 1963).

6. Herman Kahn, *On Escalation* (London: Pall Mall Press, 1965).

7. Bernard Brodie, *Escalation and the Nuclear Option* (Princeton, N.J.: Princeton University Press, 1966).

8. F. M. Sallagar, "The Road to Total War: Escalation in World War II, R-465-PR" (Santa Monica, Calif.: RAND, 1969).

9. For a later survey of the literature on limited war and escalation,

see chapter 1 of Yaacov Bar-Siman-Tov, *The Israeli–Egyptian War of Attrition, 1969–1979* (New York: Columbia University Press, 1980).

10. Ikle, *World Politics*, p. 692.

11. John Garnett, in introduction to *Theories of Peace and Security* (London: Macmillan, 1970), p. 28.

12. Karl von Clausewitz, *On War*. Edited and translated by Michael Howard and Peter Paret (Princeton, N.J.: Princeton University Press, 1976), p. 15.

> . . . war is an act of force, and there is no logical limit to the application of that force. Each side, therefore, compels its opponent to follow suit; a reciprocal action is started which must lead, in theory, to extremes.

13. See for example, Robert Osgood, *Limited War: The Challenge to American Strategy* (Chicago: University of Chicago Press, 1957).

14. This debate is covered in my *Evolution of Nuclear Strategy* (London: Macmillan, 1981), section 3.

15. The basic ideas connected with graduated deterrence were most enthusiastically propounded by Rear Admiral Anthony Buzzard, following his retirement from the Royal Navy. The term itself appears to have originated with a widely circulated memorandum by Captain Basil Liddell Hart that served as the basis for a letter to the London *Times* on August 29, 1955, which spoke of a need for a "policy of graduated deterrence" and "a plan of graduated action, based on the principle of applying the minimum force necessary to repel any particular aggression and deter its extension." Buzzard picked up the term; beforehand he was using terms like measured. See Captain Basil Liddell Hart's correspondence with Rear Admiral Buzzard, File 1/140/1, Liddell Hart Centre for Military Archives, King's College, London. Rear Admiral Anthony Buzzard, "Massive Retaliation and Graduated Deterrence," *World Politics*, January 1956. The idea was to establish an intermediate level to massive retaliation, although it was still envisaged that this graduated response would involve nuclear weapons. For a clear exposition see Richard Goold-Adams et al., *On Limiting Atomic War* (London: Royal Institute of International Affairs, 1956).

16. Consider for example this passage from William Kaufmann, writing in 1955:

> War, in fact, is a process so dynamic that it positively invites the resort to increasingly destructive expedients. Even where the parties to a conflict may want to limit the struggle, events can occur which will radically alter their calculations; and where the capabilities for virtually unlimited destruction exist, these changes may lead to the most disastrous consequences. . . . Because of its competitive character war places a heavy premium upon the attainment of an advantage, however fleeting; and this in turn invites imitation. As the belligerent strive to gain a competitive advantage, the conflict undergoes an expansion.

William W. Kaufmann, "Limited Warfare," in *Military Power and National*

Security, ed. William W. Kaufmann (Princeton, N.J.: Princeton University Press, 1956), p. 112.

Bernard Brodie's *Strategy in the Missile Age* (1959) contains no mention of escalation, but we do find the following:

> the danger that limited wars may develop explosively into total wars leads one to consider whether the net effect of readiness to adopt limited war strategies is to increase the probability of total war.

Bernard Brodie, *Strategy in the Missile Age* (Princeton, N.J.: Princeton University Press, 1959), p. 349.

17. See note 15 above.

18. Kahn, *On Escalation*, p. 3.

19. Thomas Schelling, *The Strategy of Conflict* (New York: Oxford University Press, 1960), pp. 193–4, 208. In an appendix, he refers at one point to "a gradually expanding war" (p. 259).

20. Those polled were members of the Senate Foreign Relations and Armed Services Committee, the House Foreign Affairs and Armed Services Committee, the Gaither Panel, the Military Rockefeller Panel, and a small number of other selected persons. John B. Phelps, et al., *Accidental War: Some Dangers in the 1960's* (Columbus, Oh.: Mershon National Security Program, June 1960).

21. Escalator itself was derived from *escalade*, which refers to the act of scaling, especially a fortification. Escalator began as a trade name; by 1949 it had been deemed too familiar to be subject to copyright.

22. "Escalation means the building of bigger battleships when other nations do so." *Kansas City Star*, March 29, 1938. Cited in *Supplement to Shorter Oxford English Dictionary*, vol. I (Oxford: Oxford University Press, 1972).

23. This was in the context of a debate in the United Kingdom on the wisdom of plans to fight with tactical atomic weapons. The debate is covered at length in John Groom, *British Thinking on Nuclear Weapons* (London: Frances Pinter, 1974), chapter 13.

24. Wayland Young, *Strategy for Survival: First Steps in Nuclear Disarmament* (London: Penguin Books, 1959). All emphases in original.

25. *Survival* I:1, November–December 1959. It is of note that Frankland used the term in a much broader sense than Young, as a tendency in all wars which he saw as a function of "the issues involved in the war and the passions released by it." These factors had to be taken into account in addition to "the nature of weapons in stock at the outset." It was the narrower use of Young's that became popular, in the first instance, although eventually Frankland's broader use prevailed.

26. Letter to the author, November 21, 1985. In his *Nuclear Weapons and Foreign Policy* (New York: Harper & Row, 1957), Henry Kissinger had concentrated on how to contain a limited war, using such phrases as "expansion" or "spread" to describe the pressure to be opposed.

27. "Thoughts on British Defence Policy," *The New Statesman* (December 5, 1959). Reprinted in P. M. S. Blackett, *Studies of War* (London: Oliver & Boyd, 1962), p. 84. The *Supplement to the Oxford English Dictionary* also contains a reference to the *Manchester Guardian*, November 12, 1959.

28. An example of the term's spread is Helmut Schmidt's *Defence or Retaliation* (London: Oliver & Boyd, 1962), which was first published in Germany in 1961. Escalation is used in the index for all references to the probable transformation of a limited nuclear war into a total one. The term is also used in this way in the classic Soviet work, *Soviet Military Strategy*, first published in 1962. The use was not altered in the 1963 and 1968 editions. The Soviet writers put themselves clearly on the side of those who believed that escalation from limited to general war was extremely likely and maybe even "inevitable." See V. D. Sokolovsky, *Soviet Military Strategy*, 3d ed., edited with an additional commentary by Harriet Fast Scott (London: Macdonald's & Janes, 1975), p. 170.

29. See Alex George et al., *The Limits of Coercive Diplomacy*, (Boston: Little, Brown & Co., 1971).

30. Henry Kissinger, *Necessity for Choice* (New York: Harper & Row, 1961), pp. 51, 62, 68, 82. This particular essay first appeared in *Daedalus*, vol. 89, No. 4 (1960).

31. Kaplan describes *On Thermonuclear War* "as if a giant vacuum cleaner had swept through the corridors of RAND, sucking up every idea, concept, metaphor and calculation that anyone in the strategic community had conjured up over the previous decade." Fred Kaplan, *The Wizards of Armageddon* (New York: Simon and Schuster, 1983), p. 227. Kahn did not use the term in a preview article for his book in 1959. (Herman Kahn, "The Nature and Feasibility of War and Deterrence," *SRI Journal*, 4th quarter 1959. There is a mention on p. 128 of the danger that war might "erupt" into general war.

32. Herman Kahn, *On Thermonuclear War* (Princeton, N.J.: Princeton University Press, 1960), p. 226.

33. Ibid., p. 139. This argument is often made by people who wish to escalate a little themselves, but somehow feel that the other side would not be willing to go "one step further."

34. Ibid., p. 175.

35. Glenn Snyder, *Deterrence and Defense* (Princeton, N.J.: Princeton University Press, 1961), pp. 139–140, 146, 213. He also describes a

> "spiralling" situation born of deliberate decisions to step up the intensity of the war just a bit to convince the enemy of the high costs that will follow his continued rejection of terms of settlement.

In another book of the same period, David Singer refers to "escalation from limited to general war" as a great threat. *Deterrence, Arms Control and Disarmament* (Columbus: Ohio State University Press, 1962), pp. 95–7.

36. Kahn first used the idea of the "ladder of escalation" in 1962, but it was only really publicized with the publication of *Escalation: Metaphors and Scenarios* in 1965 (see below, pp. 127–129). Kaufmann, writing in 1964, indicates that the phrase had already become current in the strategic studies community.

37. William Kaufmann, *The McNamara Strategy* (New York: Harper & Row, 1964), pp. 72, 76, 82, 83, 129. For an example of this, see Secretary McNamara's recently declassified remarks to the NATO Ministerial Meet-

ing of May 5, 1962, which led to the famous Ann Arbor speech of July 1962. Having posed the issue of "the point at which NATO, not the Soviets, would wish to escalate a non-nuclear conflict" (p. 12), he then pointed out the problems attendant on initiating nuclear war:

> As we understand the dynamics of nuclear warfare, we believe that a local nuclear engagement would do grave damage to Europe, be militarily ineffective, and would probably expand very quickly into general nuclear war. (p. 16)

However, the use of the verb "to escalate" is of note as it does recognize a degree of human responsibility for the process.

38. John T. McNaughton, Address before the International Arms Control Symposium, University of Michigan, Ann Arbor, Michigan, December 19, 1962. Cited in Kaufmann, *The McNamara Strategy*, pp. 138–47.

39. Testimony to US Senate Armed Services Committee, *Military Procurement Authorization: Fiscal Year 1964*, February 10, 1963. Reprinted in *Problems of National Strategy*, ed. Henry Kissinger (New York: Praeger, 1965).

40. Alain Enthoven and Wayne Smith, *How Much Is Enough? Shaping the Defense Program, 1961–69* (New York: Harper & Row, 1971), pp. 126–7. They consider "the risks of escalation to a general nuclear war during a tactical nuclear conflict."

41. Theodore Sorenson, *Kennedy* (London: Hodder & Stoughton, 1965), p. 513.

42. Robert Kennedy, *Thirteen Days: The Cuban Missile Crisis of October 1962* (London: Macmillan, 1969), pp. 65–6. At one point, after a U–2 aircraft had been shot down by surface-to-air missiles over Cuba, there was pressure to authorize an immediate strike against the SAM sites. The president pulled everybody back:

> It isn't the first step that concerns me . . . but both sides escalating to the fourth and fifth step—and we don't go to the sixth because there is no one around to do so. We must remind ourselves that we are embarking on a very hazardous course. (pp. 96–7)

43. Acheson, for example, suggested that following an attack on United States missiles in Turkey, the United States would be obliged to knock out a missile base in the Soviet Union. At that point, it was necessary to hope that "cooler heads will prevail, and they'll stop and talk." See Kaplan, *Wizards of Armageddon*, p. 305; Sorenson, *Kennedy*, p. 685.

44. See, for example, the interview with Admiral George Anderson, the Chief of Naval Operations at the time, *Washington Quarterly* V:4, Autumn 1982, p. 85.

> I personally believe that if an invasion had been accompanied by diplomatic assurances from the United States that its operations were directed only against Castro, then the Soviet Union would not have resorted to war.

45. Marc Trachenberg, "The Influence of Nuclear Weapons in the Cuban Missile Crisis," *International Security*, Summer 1985, p. 156.

46. In a letter sent to Khrushchev on October 22, the president said that he assumed that Khrushchev understood United States will and determination and would not therefore

> deliberately plunge the world into war in which it was crystal clear no country could win and which could only result in catastrophic consequences to the whole world, including the aggressor.

In the more personal and softer first letter sent by Khrushchev to Kennedy (the second letter was tougher in tone and more formal, apparently reflecting the views of the Soviet leadership as a whole), there are graphic images of the consequences of a failure to solve the crisis:

> If people do not show wisdom, then in the final analysis they will come to a clash, like blind moles, and then reciprocal extermination will begin. . . . we and you ought not to pull on the ends of the rope in which you have tied the knot of war, because the more the two of us pull, the tighter that knot will be tied. And a moment may come when that knot will be tied so tight that even he who tied it will not have the strength to untie it, and then it will be necessary to cut it, and what that would mean is not for me to explain to you, because you yourself understand perfectly of what terrible forces our countries dispose.

When accepting Mr. Khrushchev's letter of retreat of October 28, 1962, Kennedy noted that: "Developments were approaching a point where events could have become unmanageable." Kennedy, *Thirteen Days*, pp. 80, 89, 182.

47. Trachenberg, "The Influence of Nuclear Weapons," p. 162.

48. "Excerpts from Off-the-Record Meeting on Cuba, October 16, 1962, 6:30–7:55 p.m.," *International Security*, Summer 1965, pp. 182, 183, 193.

49. Sorenson, *Kennedy*, p. 688. Sorenson describes its advantages:

> Precisely because it was a limited, low-level action . . . the blockade had the advantage of permitting a more controlled escalation on our part, gradual or rapid as the situation required. It could serve as an unmistakable but not sudden or humiliating warning to Khrushchev of what we expected from him. Its prudence, its avoidance of casualties and its avoidance of attacking Cuban soil would make it more appealing to other nations than an air strike, permitting OAS and Allied support for our initial position, and making that support more likely for whatever air-strike or other action was later necessary.

50. Summary Record of NSC Executive Committee Meeting, No. 8, October 27, 1962, 4:00 P.M., *International Security*, Summer 1985, p. 203.

51. Kennedy, *Thirteen Days*, p. 109, Kennedy told Dobrynin that this was a "statement of fact" not an "ultimatum." This analysis very much follows that set out by Alexander George in his chapter on Cuba in *The Limits of Coercive Diplomacy*.

52. See George, *Limits of Coercive Diplomacy*, p. x.

53. George, Ibid., p. 18. George defines this strategy as follows:

> Force is used in an exemplary, demonstrative manner, in discrete

and controlled increments, to induce the opponent to revise his calculations and agree to a mutually acceptable termination of the conflict.

54. Thomas Schelling, *Arms and Influence* (New Haven, Conn.: Yale University Press, 1966), p. 166.

55. Albert and Roberta Wohlstetter, "Controlling the Risks in Cuba," Adelphi Paper No. 17, (London: ISS, February 1965). Elsewhere they refer to an "ascending series of non-nuclear threats and actions." (pp. 16, 18, 19)

56. Talk at RAND, May 1963, quoted by Barry H. Steiner in *Bernard Brodie and the American Study of Nuclear Strategy* (Unpublished Paper: California State University), p. 338. As time passed he became nervous that this lesson had not been learned:

> An interesting and possibly alarming aspect of the Cuban Crisis of 1962 was the degree to which the crisis stimulated even among American Administration leaders a tendency to think or at least talk in such simplistic but absolute terms, despite the sophistication that had presumably been accumulating in the preceding months concerning the appropriateness of flexible response and the feasibility of limited options.

Brodie, *Escalation and the Nuclear Option*, p. 118.

57. Ibid., p. 353.

58. Quoted in Steiner, p. 302. Steiner quotes two generalizations from this paper (pp. 299–300):

> new and higher levels of violence introduced by one side in a conflict is highly likely to be met by the other side, so long as the latter possesses the appropriate capabilities and is not ready to relinquish its position. . . . levels of violence tend almost always to move upwards, rarely diminishing, until the actual ending of conflict.

59. Brodie, *Escalation and the Nuclear Option*, pp. 32, 113.

60. Ibid., pp. 34, 43, 57, 71, 88, 118. Brodie's assumption was that early use of a few nuclear weapons was less likely to push things out of control than intensive and prolonged nonnuclear warfare. (For a similar suggestion see Kissinger, *Necessity for Choice*; see p. 116.)

61. Klaus Knorr and Thornton Read, *Limited Strategic War* (New York: Praeger, 1962). Other authors expressed concern about the pressures from public opinion in favor of escalation (pp. 15, 18, 76, 127, 176, 214, 223).

62. "Limited Retaliation as a Bargaining Process," in Knorr and Read, *Limited Strategic War*, pp. 149, 157. In an earlier version of this paper Kaplan had referred to "a reprisal process" rather than escalation. Morton A. Kaplan, *The Strategy of Limited Retaliation*, Policy Memorandum No. 19 (Center of International Studies: Woodrow Wilson School of Public and International Affairs, April 1959), p. 13.

63. Herman Kahn, "Some Comments on Controlled War," in Knorr and Read, *Limited Strategic War*, pp. 38, 40, 41, 64.

64. Herman Kahn, *Thinking the Unthinkable* (New York: Horizon Press, 1962), p. 47.

65. Ibid., p. 185.

66. Ibid., p. 186.

> Roughly, the higher one is on the Ladder the more intense the dispute, and the closer one is likely to feel to large-scale violence. There are of course many other possible sequences of escalation. Moreover, even on the above Ladder many of the rungs can be skipped. . . . Nor is the order sacred, and many people would interchange some terms. There is also no implication that one must go inexorably up the Ladder. One could go either up or down step by step or skip steps. Finally, and most importantly, the Ladder is intended to describe only a class of situations.

67. Ibid., p. 196.

68. Kahn, *On Escalation*, pp. 281, 290.

69. Luigi Nono, the modernist Italian composer, used Kahn's ladder as the theme for a musical composition, which he dedicated to the National Liberation Front of Vietnam, which moved from (1) "crisi manifesta" to (44) "spasmo o guerra insensata." Ikle, *World Politics*, July 1967, p. 693.

70. This point about thresholds was by no means original to Kahn. Again the same idea can be found in Schelling, *Arms and Influence*, pp. 164–5.

> It is undoubtedly in the interest of limiting war that some obvious firebreaks and thresholds occur. There are subclasses or patterns or conventional boundaries to help find a stopping place. It is hard to stop without an obvious stopping place . . . some thresholds have made the claim to being the "ultimate limit", the last stopping place before all-out war.

71. Halperin, *Limited War in the Nuclear Age*, pp. 14, 31, 37. These limits (by and large derived from Korea and reinforced subsequently) were:

> the nonconfrontation in battle of American and Soviet troops; the confinement of a theater of war to a single country; the recognition of sanctuaries of supply lines and training zones outside the battlefield area; and the non-use of nuclear weapons.

72. Morton Halperin, *Contemporary Military Strategy*, pp. 92–3.

73. Kahn, *On Escalation*, pp. 6–7, fn. 5.

74. Schelling, *Arms and Influence*, p. 173.

75. See, for example, William E. Simons, "The Vietnam Intervention, 1964–65," in George et al., *The Limits of Coercive Diplomacy*, p. 148.

76. Robert Osgood, "Reappraisal of Limited War," in *Problems of Modern Strategy* ed. Alastair Buchan (London: Chatto & Windus, 1970), p. 109. See also his *Limited War Revisited* (Boulder, Colo.: Westview Press, 1979).

77. Stephen Rosen, "Vietnam and the American Theory of Limited War," *International Security*, Fall 1982.

78. For an early version of this argument see Colonel A. P. Sights, Jr, "Graduated Pressure in Theory and Practice," *US Naval Institute Proceed-*

ings, July 1970. See the sources gathered together in support of this thesis by Colonel Harry Summers, *On Strategy: A Critical Analysis of the Vietnam War* (Novato, Calif.: Presidio Press, 1982), especially pp. 116–24.

79. See, for example, R. B. Smith's description of a draft memorandum by William Bundy of March 1, 1964, in which he developed the case for employing military threats to coerce North Vietnam. While Bundy was well aware of the differences between this situation and the missile crisis, he was nevertheless thinking in terms of trying to repeat the "miracle of the Cuba crisis." R. B. Smith, *An International History of the Vietnam War*, vol. II, *The Struggle for South-East Asia, 1961–65* (London: Macmillan, 1985), p. 244.

80. See, for example, Smith in note 79, pp. 254–8; as well as George et al., *The Limits of Coercive Diplomacy*, on which Smith draws.

81. Kaplan, *Wizards of Armageddon*, pp. 332–6. According to Kaplan, when considering the bombing campaign against the North in 1965, McNaughton asked Schelling how to devise a plan to send the appropriate signals to Hanoi. Kaplan comments, with some satisfaction:

> So assured, at times glibly so, when writing about sending signals with force, inflicting pain to make an opponent behave and weaving patterns of communication through tactics of coercive warfare in theory, Tom Schelling, when faced with a real-life "limited war," was stumped, had no idea where to begin.

In Schelling's defense, it could be pointed out that his theories had not been developed for this sort of occasion and, in fact, though this is not clear from Kaplan's account, he was highly skeptical as to whether his theories could apply in this situation.

82. Kahn, *On Escalation*, p. 112.

83. Schelling, *Arms and Influence*, pp. vii, 84, 85, 166, 171–2.

84. Professor Robert H. Johnson, letter to the author, December 5, 1985.

85. *Pentagon Papers*, vol. III. Bundy memo, February 10, 1964, pp. 312–5; Taylor and Sharp, pp. 315–8; McNaughton, pp. 694–702.

86. In a memo in the summer of 1965, advocating a continuation of the bombing program, McNamara laid down the principles which should govern it, which largely emphasized the need to maximize its political impact. The final requirement was that the program should:

> *Avoid undue risks and costs.* The program should avoid bombing which runs a high risk of escalation into war with the Soviets or China and which is likely to appall allies and friends.

Pentagon Papers, vol. IV, pp. 28–9. Guenter Lewy argues that "the decision for 'gradualism' was made primarily because of fear of Chinese intervention." "Some Politico–Military Lessons of the Vietnam War," *Parameters* XIV: 1, Spring 1984, p. 9.

87. William Simon in George et al., *The Limits of Coercive Diplomacy*, p. 197.

88. Department of State, Policy Planning Council, *Alternatives for the Imposition of Measured Pressures against North Vietnam*, March 1, 1964. See Robert H. Johnson (the director of the study), "Escalation Then and Now," *Foreign*

Policy, Fall 1985. Also David Halberstam, *The Best and the Brightest* (New York: Random House, 1972), pp. 434–9.

89. George Ball, cited in Leslie Gelb and Richard Betts, *The Irony of Vietnam: The System Worked* (Washington, D.C.: Brookings, 1979), p. 111. Senator Mike Mansfield is also cited as opposing further commitment at the same time with the words, "Remember, escalation begets escalation" (p. 125).

90. In addition to George's, *The Limits of Coercive Diplomacy*, which was not finished until he had moved to Stanford from RAND, see also Sallagar's, *The Road to Total War: Escalation in World War II* (Santa Monica: RAND, April 1969), and Ikle's, *Every War Must End*.

91. George, *The Limits of Coercive Diplomacy*, p. 216.

92. Rosen, "Vietnam and the American Theory of Limited War," p. 83.

93. Consider, for example, the following from Enthoven and Smith, *How Much Is Enough*, p. 127.

> The rules of a limited nuclear war would be complex and unclear, even assuming that all parties involved tried to abide by the rules.
>
> In conventional war the rules are comparatively simple and clear. The upper limit of escalation is not shrouded in uncertainty. Conventional war can still be a test of strength; tactical nuclear war, because of the ambiguities mentioned, is more likely to be a test of will.

94. For an important critique of the lack of attention paid by Schelling and Kahn in their discussion of "commitment" to the real interests of the powers concerned, see Stephen Maxwell, *Rationality in Deterrence*, Adelphi Paper No. 50, (London: ISS, 1968).

95. See above, pp. 129–131.

96. See, for example, Sallagar's study for RAND, *The Road to Total War: Escalation in World War II*, pp. x–xi, which warned that

> The insidious appeal of the gradualist approach to escalation is that it could blunt the caution that would be observed in case of a major, abrupt change in the level of violence.

PART II

Soviet and American Perspectives

CHAPTER 5

Nuclear Deterrence and Technological Change: Retrospect and Prospect

COLIN S. GRAY

TECHNOLOGICAL CHANGE: PROBLEM AND CONDITION

This chapter explores the complex relations among technology, policy, strategy, and doctrine. The purpose here is to contribute to understanding and not to advocate particular propositions for policy or particular weapons.[1] Hypotheses of major interest for this discussion include the following:

1. That deterrence fundamentally is an existential condition between nuclear-armed states; the details of posture and doctrine matter little if at all.[2]
2. That technological change is a routinized feature, really a condition, of the modern world and on balance is a source of net security as opposed to insecurity.
3. That the dangers of war reside more in political than in military–technological factors.[3]
4. That military technology, although possibly less important for the preservation of peace with security than essentially political incentives and disincentives to fight (see 3 above), nonetheless can have an important benign or malign influence on the choices made by statesmen.
5. That many military problems at root are technological in character and can be alleviated or resolved only by technological developments.

These hypotheses illustrate more a framework of concern than a detailed strategy of analysis. Students of nuclear strategy notice the plain absence of the discipline of historical experience that restricts confidence in what passes for knowledge in their subject. The game aspects of theory on nuclear strategy, in which nuclear *weapons*, not infrequently, really are not treated as if they were weapons (as traditionally understood), reflect both a healthy appreciation of the uniquely horrible features of these devices and a growing, and probably unhealthy, incredulity about the whole subject.

Scholars who seek to probe the mysteries of the possible causes of the next war may find some cold comfort in the liveliness of the continuing debate among historians over the causes and courses of the general wars that actually have occurred in this century.[4] The military technologies of 1914, 1939, and today are very different, but one suspects that the capacity of professional defense planners to err in their prognoses is no less great in the 1980s than it was fifty and seventy years ago,[5] notwithstanding the apparently *absolute* characteristics of nuclear weapons. Defense planners, East and West, cannot be confident today that they either have, or even can develop, a military instrument of swift decision, given the residual punitive power even of heavily attrited nuclear-armed forces. The general staffs of Europe in the period immediately preceding 1914 should not have been confident that they had a certain formula for quick victory—be it *the Plan*, in the German case, or morale, in the French case—but their generally meticulous, almost infinitely detailed, study of modern war was framed by the lessons of 1870–71.[6]

The nuclear age, or the age in which the war between great superpower-dominated coalitions of allies and clients either would or could be waged with nuclear weapons, unsurprisingly has produced some divergent intellectual and policy responses among theorists and officials in the West.[7] The invention of nuclear weapons, married to delivery vehicles for long-range bombardment, has created a generally recognized crisis of understanding of the nature of modern war that can be resolved only by experience of the dreaded event. Objectively speaking, with the benefit of hindsight, crises of understanding of the nature of modern war are far from constituting a strictly contemporary phenomenon. The military planners of 1914 had embraced doctrines of war that soon were proven by events to reflect inadequately the meaning of the industrial revolution for national capacities to absorb loss and generate combat (though not decision) power.[8] It is distinctly possible that Western defense communities today may some day be proved by unexpected events to have taken undue account of, and undue comfort in (for deterrence), the potential for societal–terminal catastrophe that looms, beyond question, in contemporary nuclear arsenals.

The United States defense community, in common with every other defense community that is blessed or cursed with custodial roles for nuclear weapons, is obliged to think about, and plan for, the actual use of these weapons. Responsible officials are denied the professional luxury of being able to take refuge in the assertion that nuclear weapons constitute an "existential deterrent."[9] Even if one suspects that an important truth may lurk in the existential hypothesis, management and strategy questions remain for officials to answer. If one ventures into the forbidding realm of nuclear planning and worries about nuclear employment policy, one is at risk: (1) to the charge of an attempt to *conventionalize* what is very different from conventional weaponry;[10] (2) to the accusation, or perhaps observation, that, psychologically, one is denying what should be discerned as the plain fact of a true revolution in the means, and in this case even the meaning and utility, of war; and (3) somewhat generically, that one is endeavoring, inappropriately, to explain "nuclear politics with prenuclear concepts."[11] Robert Jervis, by far the most cogent purveyor of these linked charges, may be correct. But he may be wrong.[12] Moreover, Jervis's charges are not verifiable—short of the field test by battle that all doctrinal contenders agree must be prevented at many, though not all, costs.

There has long been a problem in understanding the tactical and strategic meaning of new weapon technologies, in the context of other factors that help generate or retard "fighting power"[13] (including Clausewitz's composite concept of "friction."[14]) However, save for the very imperfect analogy of poison gas in World War II, there is no historical precedent for the fundamentally irresolvable—by careful study in peacetime—uncertainties that pertain today to the character of future war. For a prominent example, there are scientific grounds for considerable doubt over the likely occurrence of a very severe *nuclear winter*, were one or more of the nuclear powers to execute targeting plans that produced, on a very large scale, the burning of urban areas and forests and the violent transportation of dust and soot particles into the upper atmosphere.[15] The scientific possibility of nuclear winter, however, is not really at issue. The interesting and policy-relevant questions have to do with: (1) the targeting choices the combatant powers would make; (2) the likelihood, or otherwise, that restraint in targeting could be exercised in the event of war; and (3) the sensitivity of the extent of climatological deterioration to changes at the margin in weapon design and nuclear employment.[16] One can assert with some confidence that no country, knowingly, is going to implement a military strategy that must be self-defeating.

It is tempting to be somewhat dismissive of the significance of technological change for conflict and stability, on the apparently reasonable grounds that incentives to compete and even to fight tend essentially to be political in character. Preferred military doctrines have a way in peacetime, of interpreting new technologies according to the dictates of organizational convenience. Because of "the hiatus of the mobile arm" ("on" and "from," but not so much "to," the battlefield), the Great Powers of Europe might have realized that their armies had ceased, for a while, pending the maturity of vehicles powered by the internal combustion engine, to be instruments capable of resolving political problems at tolerable cost.[17] For very understandable reasons, such realization escaped them. Preeminent though political factors tend to be in the history of conflict, there should be no denying that technological change affects the military capabilities of states and can affect them decisively, and that assessments of capability fuel and help direct assessments of threat. Furthermore, some particular military problems, in reality if not in theory, do not lend themselves to "workarounds" by politics or strategy. From 1914–18, there were no practically available technological solutions to the problems of the stalemate on the Western Front, just as there were no clever strategies or tactics by which the deadlock could be resolved "at a stroke."[18] Mobility on the battlefield, to a degree that would reopen the prospects for rapid decision, could be restored only by technological change on a scale that could restore the possibility of decisive maneuver. Today, the problem, if one chooses so to regard it, of long-range nuclear threat cannot be removed by arms control, disarmament, or by changes in strategic doctrine.[19] Soviet long-range weapons technology poses a threat of catastrophic, and probably fatal, dimensions to American society. This technological threat can be answered, if at all, only by United States technological change.

A later section of this essay discusses President Reagan's Strategic Defense Initiative (SDI) in some detail. Whether particular architectures of defensive weaponry are judged to be technically feasible, cost effective, and desirable in broad policy perspective, there is nothing inherently unreasonable in the proposition that the high technology dimension of threat to Western security should be met with high technology solutions. Without seeking to deny the unique, single-unit destructive power of a nuclear weapon, one should recognize that the SDI carries the potential for effecting as profound a shift, or more, in doctrines of war as did (1) the introduction of generally reliable cannon for siegecraft in the fourteenth and fifteenth centuries;[20] (2) the improvement of infantry small arms for resisting cavalry in a shock role in the nineteenth century; (3) the (new) perfection of artillery weapons and, above all, techniques between 1914 and 1918 (for the frustration of the offensive potential of infantry and later for the "unlocking" of trench systems);[21] (4) the restoration of battlefield mobility with the introduction by 1939 of tanks that were true breakthrough weapons; (5) the various impacts of airpower on land and naval warfare in World War II; (6) the introduction of fission and then fusion weapons in the mid-1940s and mid-1950s, respectively, married to delivery vehicles and doctrines borrowed from the prenuclear era (airpower);[22] (7) and the maturing of long-range ballistic missile weapons in the late 1950s.

Military history is by no means exclusively a history of technological change. Of course, "great captains," competent staffs, superior training, sheer quantitative advantages, innovative tactics, suitable strategy, high morale, and stability on the home front (*inter alia*) can be of enormous, and not infrequently decisive, significance. Nonetheless, cumulatively dramatic changes in weapons technology, or in civilian technology applicable to the waging of war,[23] have been endemic in recorded history—as has resistance to change. Settled doctrines, the lessons of "the last war," patterns of budgetary allocation, and individual careers can all be placed at risk by the prophets of military–technological change. Given the necessary technical immaturity of genuinely new weapons and the likely obscurity of the details of their proper tactical employment, the skeptical standard-bearers for the old order typically are as correct in their doubts concerning the near-term utility of new weapons as they are incorrect in their medium-term prognoses. In the still longer run, both the initial skeptics and the prophets are virtually certain to be confounded, as technical and tactical solutions, or partial solutions, are invented to offset the military potential of the change in question. However, there is no law of innovation and counterinnovation that guarantees the eventual restoration of utility to previously abandoned ways of conducting war.

Technological change is both a problem and a condition. Because it is a problem, it unsettles the old order; frequently, in effect, people endeavor to deny that it is a condition or that it is a condition pertinent to doctrinal innovation. The contemporary debate over strategic defense is reminiscent of the very bitter debates of the 1920s and 1930s over the character of future land warfare. The very recent experience of the greatest war in history yielded the view that artillery was the "break-in" weapon, that infantry held the ground taken by artillery, and that tanks were weapons for the support of the infantry. This view was not merely plausible, it was demonstrably correct with reference to 1917 and 1918. The very varied experience of 1939–45 would show that virtually all views on the utility of tanks for different missions were correct—depending on the quantity and quality of armament, the tactical skill and morale of the opposing forces, the nature and extent of the relevant local terrain, and last, but not least, the weather.[24]

In common with the prudent skeptics of the 1920s, many

opponents of the SDI (as an eventual array of weaponry, not as a research program) believe, with evident sincerity, that they understand the essential nature of contemporary war. The SDI is denounced as an effort either to repeal or to deny that essential nature. Since perfect defense is a purely nominal aspiration, recognized by all theorists and officials as a worthy but unattainable vision, the SDI, in this critical view, offers only a spurious promise of transforming the landscape of security. American society will continue to be at nuclear risk, whether or not the SDI can provide an historically unprecedented robustness in physical protection. Opponents of the fairly steady trend of the 1970s and early 1980s toward a *classical strategy* of war-fighting deterrence deny that they favor a strategy of mutual assured destruction or mutual societal vulnerability.[25] Indeed, they argue—with some plausibility—that mutual assured destruction is a condition of East–West security relations; it is not a matter subject to policy preferences.[26] Tinkering with doctrine, war plans, and posture cannot, in this perspective, provide an escape from the threat posed by nuclear weapons.

TECHNOLOGICAL CHANGE, DETERRENCE, AND THE UTILITY OF FORCE

The idea of deterrence is as old as the history of conflict. States are deterred if they believe that the costs of war outweigh the benefits. Formulated in a more useful fashion, deterrence should obtain if:

1. the political objectives sought through military action are deemed to be beyond confident reach, for reason of the uncertainty of military success;[27]
2. the costs in assets likely to be lost in war are predicted to be so high that the political objectives secured by military success would be deprived of much of their meaning.

The nuclear fact in the modern world has increased stability in that the threshold of political offense for the resort to force among nuclear-armed states undoubtedly has been raised.[28] Incentives to alleviate or resolve political problems by force on the part of nuclear-armed states in their mutual dealings have been more than counterbalanced by the disincentives provided by the expectation, or even just the possibility, of triggering nuclear war. A somewhat comforting, even complacent, view of the seeming "meta-stability" imposed by the nuclear condition, has some commonsense appeal.[29] Unfortunately, the belief that the very existence of nuclear weapons suffices to deter both direct armed conflict between the superpowers, and other conflicts that Moscow and Washington believe might escalate to a direct struggle between them, presumes a modesty of political motives and a manageability of sequences of political and military events that history, as well as a generic prudence, suggests to be unwise. Given how destructive a nuclear war could be, reasonable, self-regarding leaders who contemplate the possible, as well as just the more probable, consequences of their actions should always be deterred from engaging in conflict that could become nuclear. But, the very destructive potential of nuclear weapons that works to discourage adventurous diplomacy for fear of possibly uncontrollable consequences also can serve both as a uniquely promising tool of political intimidation and perhaps as a very efficient instrument for rapid military decision.

A major problem for defense planners is the need to identify criteria for assessing the adequacy of a deterrent posture. Who is to be deterred from doing what? How strong a political motivation to take the inimical action in question should be presumed? And what is the prospect that should have the suitably deterring effect? The belief that governments should regard nuclear arsenals as if they constituted a holocaust machine is not difficult to comprehend, but it has no merit as operational advice. Whether nuclear weapons could function truly as weapons, capable of implementing strategy in pursuit of political goals or war aims worthy of the name, is an interesting question in its own right; but this subject, alas, is not strictly an academic one. For good or ill, the Western Alliance, with its adherence to the concept of flexible response, is heavily dependent on the credibility of nuclear threats.[30] The details of the threats posed have altered as the strategic balance has evolved and as doctrine has shifted, but the strategic missions of United States nuclear forces have not altered in any notable way during the past thirty-five years. The most burdensome of these missions, certainly the one that poses the greatest challenge to the planners of force structure, is the requirement that United States strategic nuclear forces extend deterrence over geographically distant allies and impose a suitable weight of escalation discipline on the enemies of United States expeditionary forces fighting far from home.[31]

The problem of the United States in nuclear deterrence is not the need to persuade an enemy that the United States would retaliate in the event of a nuclear assault on its territory; rather, the problem is the continuing need to persuade an enemy that the United States would, or just might, have resort to first use of nuclear weapons—including so-called strategic nuclear weapons (or central systems, in the jargon)—for the purpose of restoring deterrence should conventional defenses be overrun. Extended deterrence is a heroic mission for a country incapable of a preclusive offensive counterforce strike and that, in addition, is almost totally bereft of defensive counterforce capability. It is one thing to deter an enemy from initiating a nuclear attack; it is quite another to avoid self-deterrence if the burden of nuclear escalation is on your shoulders. That is the essence of the United States deterrence problem; it flows from the geography of East–West conflict and the character of Western European societies that decline to provide a truly robust, local nonnuclear defense effort.

In 1959, Albert Wohlstetter argued that the balance of terror was a delicate one, requiring constant attention in order to avoid the dangers of a technical instability deemed to lurk in theoretically vulnerable retaliatory forces.[32] It is probably sensible to assume that the balance of terror is delicate even though there are ample grounds for suspecting that that is not the case. It may be more sensible to argue that the balance of terror is massively indelicate to technical perturbation all the time the political incentive to fight is low. All the while Soviet leaders are not interested seriously in enquiring as to the availability of military solutions to the political problems that beset their insecure territorial and hegemonic empire, theoretical windows of military opportunity may come and go without much consequence.[33] The key issue is how much deterrence is required.

Day in and day out, year after year, there is unlikely to be even a single occasion when a Soviet leader asks, "Am I deterred from starting a war directly with the United States?" Existential deterrence indeed does obtain. But, even though Western nuclear forces serve as an essential backcloth to peacetime diplomacy and contribute vitally to negative assessments as Soviet officials wrestle with the mysteries of "the correlation of forces," the details of their quality and quantity truly may make the difference between war and peace only once in thirty or forty years.

Nearly a quarter-century after the Cuban missile crisis, Western defense communities are wont to forget that acute, war-threatening, international crises can occur. The existential deterrent judged adequate to discourage sober and rational Soviet leaders from seriously contemplating taking the highest of risks in quest of positive advantage would likely begin to appear only marginally competent to dissuade truly desperate Soviet leaders confronting what they might define as a crisis of empire and regime. One may believe that Soviet leaders will *never* confront a political crisis: (1) of regime-threatening severity, and (2) to which a military strike westward would be believed to constitute an arguably appropriate response, or component of response. But never is a long time; the Soviet *imperium* at home and abroad contains growing and disparate amounts of fissionable fuel;[34]

and, in general, it is imprudent to rest one's country's security on the best case analysis of a Soviet Union eternally easy to deter.

This author has been criticized, by Michael Howard among others, for advocating what would or might amount to a United States–NATO ability not merely to deter nuclear war but also to win nuclear war.[35] This advocacy has been misunderstood, though, admittedly, it does lend itself to misunderstanding.[36] Victory, as Harry Summers notes in his rigorous analysis of strategy in Vietnam, need not be defined in terms of the total military overthrow of the enemy.[37] When Secretary of Defense Caspar Weinberger refers to the contingent United States intention to employ force "to seek the earliest termination of conflict on terms favorable to the United States, its allies, and its national security objectives,"[38] he means that the United States and its allies would intend to prevail by way of achieving their political objectives in war.[39] Those objectives, via a military determination to defeat the enemy's strategy, would, in the first and possibly only instance, be confined to repulsing attack and restoring, at a minimum, the territorial status quo ante. Given the strong likelihood that United States and NATO conventional and theater nuclear forces would be inadequate to prevail even in that relatively modest endeavor, strategic nuclear planning must embrace some theory of what Thomas Schelling termed "compellence."[40] United States "central systems" are required to be able to repeal the verdict of the land battle. To be polite, that is a difficult mission. A Soviet Union sufficiently politically motivated to attack in the first place, sitting on a growing quantity of probably hard-won gains in Europe, and prepared culturally for great sacrifice of life and property on the part of civilians, could take a great deal of "compelling."

Whatever the fashionable trend in strategic doctrine may be, technology and program action (or inaction) on both sides ultimately dictate what can be done. Second-strike retaliation against urban–industrial targets enjoys high credibility at the levels of policy and theory and even enjoys a high credibility in prospective mission completion; hence, it might seem to be *the* feasible option. Unfortunately, such second-strike retaliation, though relatively technically undemanding once launched, could be difficult to execute (command and control are guaranteed only if one strikes first) and would accomplish nothing of political value, even if it were competently conducted.

The question of the adequacy of the diverse nuclear deterrent of the West cannot be approached save with reference to the structure of Western deterrence policy as a whole. Deterring nuclear war is important, but it is not as important as the mission of deterring war per se—because nuclear war is likely to emerge out of a nonnuclear war. Stability in the central strategic relationship between the superpowers can translate into instability overall.[41] Notwithstanding the worthy aspirations expressed in recent years for the fielding of new emerging technologies to enhance conventional deterrence, historical experience, policy preferences in Western Europe, and the logic of strategic geography suggests that, pending a new approach to strategy (cited below), the United States requires a measure of military advantage in strategic forces over the Soviet Union in order for NATO to enjoy the benefits of a tolerably even balance of military power.[42] Such a military advantage, which really is a strategic requirement for NATO security under the present policy, can be secured and sustained only by the timely exploitation of new technologies. Flexible response strictly may not require the capability for what Herman Kahn termed "escalation dominance,"[43] but it certainly requires that the United States be able to initiate a central war armed with forces capable of imposing escalation discipline on the Soviet Union, for the purpose of restoring deterrence (as only the most minimal of objectives).

Looking to the future, and bearing in mind the possibility of new strategic defenses being deployed on both sides, it would be prudent for the West to cease thinking of United States strategic nuclear forces as the extended deterrent for the last-ditch protection of peripheral Eurasia. Instead, the extended deterrent should be the mobilizable defense potential of the United States–led coalition, knit together by Western maritime preponderance. If NATO could deny short-war success to Soviet arms, it and its Asian allies could threaten the Soviet state with a protracted war against a coalition that enjoyed on the order of a four or more to one advantage in economic strength.

Changes in technology by both superpowers have promoted reconsideration of the utility of offensive counterforce in the United States and should have a similar effect in the Soviet Union. Strategic ideas and the policy guidance for force development that flows from them—albeit typically in somewhat ambiguous and rough-hewn form—have their season in technical feasibility. Mutual assured destruction, insofar as it reflected probable military reality in the late 1960s, was not then so much a matter of policy preference as rather an expression of contemporary counterforce incompetence. Similarly, the clear trend toward a partial war-fighting theory of deterrence, identifiable in NSDM–242 through the current policy guidance, has found its inspiration and justification in part, though only in part, in the prediction and achievement of MIRVed missile accuracy approaching and now far surpassing 0.1 nm.[44] Just as MAD as a strategic concept was overtaken by technological developments, so the technical integrity of the "victory denial" leitmotiv of President Carter's "countervailing strategy," and President Reagan's "prevailing strategy," is imperiled by technological change in the most important target sets of both superpowers.

The plainest statements explaining the victory denial theory of deterrence were provided in April 1983 in the first Report of the Scowcroft Commission. The commissioners claimed that "deterrence . . . requires military effectiveness"[45] and that:

> We must be able to put at risk those types of Soviet targets—including hardened ones such as military command bunkers and facilities, missile silos, nuclear weapons and other storage, and the rest—which the Soviet leaders have given every indication by their actions they value most, and which constitute their tools of control and power.[46]

As in the 1960s, when ICBM silo construction and rapid proliferation and *Yankee* class SSBN deployment reduced dramatically the counterforce value of the United States strategic forces, so in the late 1980s and 1990s, again, military effectiveness promises to be increasingly difficult to achieve. First, consider the more important actual and predictable features of the Soviet problem. The United States:

1. ICBM force, in silos, certainly could be launched in a prompt second-use mode on confirmation of attack.[47] (The modernization of ground, air, and space-based C^3I provides a new quality of prompt responsiveness to attack. Precursor assault against key elements of the C^3I architecture would serve to provide warning.)
2. ICBM force may be housed in superhard silos, may be actively defended, and may have a mobile component after 1992.
3. SSBN force is acquiring more sea room while still on station, with the deployment of the *Ohio* Class equipped, after 1989, with the *Trident* D–5 SLBM.
4. Bomber force is being modernized with the B–1B, which has a more rapid takeoff than the B–52, can use shorter runways, and is comprehensively hardened against nuclear weapon effects.

The United States strategic forces' triad, considered synergistically as a complex target set for Soviet war planners, is evolving in ways that should be profoundly discouraging. Stated directly, a Soviet preemptive counterforce attack, particularly, though not only, considered in a plausible operational context of international crisis or war (that is to say, with United States strategic forces in a condition of "generated" alert at DefCon 1), is likely to fail catastrophically. Second, consider the parallel United States problem. The Soviet Union:

1. Has increased the blast resistance of its more modern silos more than threefold over the past decade, has a true superhardening option available should it so choose, and plainly intends to deploy a large fraction of its new fifth-generation systems (e.g., the SS–24 and the SS–25) in a mobile (rail and truck) mode.[48] In common with the United States, the Soviet Union appears to have acquired the quality of C^3I suitable for a launch-under-attack firing tactic.
2. At last, and inevitably, has developed a much quieter propulsion system for its new class of SSBN (the *Delta* IV), and has evolved new deployment patterns in the far North—both of which, married to range growth for SLBMs, render the United States strategic ASW mission increasingly stressful (though not necessarily hopeless, by any means).[49]
3. Has developed a new bomber and new long-range ALCMs that may serve to effect a massive upgrade to the air-breathing threat to a United States that has no air defense.
4. Has protected its leadership cadres by superhard structures and by the proliferation of semihard bunkers.
5. Would certainly move its ground forces out of harm's way "into the field," leaving United States weapons with empty barracks to hold at risk.

The details above are intended, nonexhaustively, to illustrate a growing problem for the strategic policy of both superpowers. Some liberal critics of the Reagan and late–Carter administrations have been very free with their criticism of the war-fighting trend in the official theory of deterrence—as expressed, reportedly, in PD–59 and NSDD–13, and very visibly in the language of the Scowcroft Report and in such weapon programs as the MX ICBM and the *Trident* D–5 SLBM. But generally, those critics have failed to notice that evolution in the Soviet target base has been functioning, nonaccidentally of course, to thwart in practice the efficacy of that United States theory. However, pending technological development in the realm of defensive weapons, the United States has no sensible choice other than to seek military effectiveness through offensive counterforce.

The societal punishment threat of assured destruction, in its several variants, comes as a gift, courtesy of strategic forces designed to execute technically more demanding missions. Massive counter-urban threats are of no military relevance to the actual conduct of central strategic war, but whatever deterrent value flows from the United States residual capability to blow down or burn Soviet cities is extant, whether United States officials want it or not. It is probably worth mentioning, for the benefit of those like Robert Jervis who still believe that "MAD is the best possible deterrence,"[50] that one of the few items on which everyone who has studied nuclear winter phenomena agrees, is that the targeting of cities is a uniquely poor idea.[51]

The technology and tactics of passive protection appear to be winning the race with the evolving lethality of more and more accurate offensive weapons. Cruise missiles that can strike within one hundred feet of their aim point and ICBMs and SLBMs with CEPs of close to three hundred feet can only be effective if their target intelligence is as good as their aim. Target sets critical to the countervailing strategy increasingly are protected deep underground, by proliferation, by agility, and—predictably—by more and more active defense. The Soviet Union is not going to abandon its theory of war simply because the heavy artillery of its ICBM force cannot "decapitate" United States command and control, or disrupt and massively disable United States strategic forces. Western policy makers, seemingly focused almost exclusively on the United States Strategic Defense Initiative, should appreciate the longevity and seriousness of the Soviet commitment to a balanced offense–defense posture.[52]

For the United States, the receding accessibility of the Soviet state assets critical to the official theory of deterrence should, for the time being at least, be viewed as a technical challenge to be addressed energetically. Mobile missiles are not invulnerable; rather, they are difficult to find and strike in a timely manner. Similarly, Soviet ground forces do not achieve inherent immunity by moving out of their casernes. Even if the prospects for offensive counterforce competence are better than has been suggested here, the Western Alliance has very pressing reasons to explore the feasibility of an important change in the terms of deterrence to be triggered by a new efficacy in strategic defense. This subject is discussed later in some detail. At the most general of levels, the United States and its allies have to be seriously interested in defense because:

1. The Soviet Union is pursuing the full range of defensive technologies with great vigor and, one day, may be in a position to implement a true *classical strategy*.
2. Sometime in the future, the deterrence system may fail, or fail to function altogether.
3. The offense-dominant strategic policy of the West today affronts the humane values it seeks to protect—witness the fact that tolerably honest public discussion of United States and NATO nuclear strategy is profoundly politically inexpedient (in order to justify a new weapon like the MX, which is mandated by U.S. strategic policy guidance, officials are obliged to seek cover in the asserted needs of the bargaining process in arms control).
4. Difficult though this is to discuss outside the ranks of defense

professionals, strategic defenses could make a dramatic difference to the amount of suffering that would attend a major war. In short, the second of the traditional goals of arms control—that damage in war should be minimized—should not be eschewed peremptorily in fear that its pursuit might harm achievement of the first goal, that the danger of war occurring be reduced.

5. Last, but by no means least in importance, in the light of emerging technology trends in offense and defense and in view of what the Soviet Union has revealed in its recent and distant history concerning its preferred mode for the conduct of war,[53] there are grounds for believing that strategic defensive technologies may provide the basis for a more robust system of deterrence.

OFFENSE, DEFENSE, AND THE FUTURE OF DETERRENCE

Some plainly inaccurate statements have been offered in criticism of President Reagan's SDI. The fact that some no less accurate, and many simply confused, statements have been tendered in attempted justification of the Initiative does not excuse the former. Lawrence Freedman, for a prominent example, has invented his own demonstrably false version of the SDI and has the gall to charge the United States government with effecting a "retreat" from the president's speech of March 23, 1983, in favor of the "intermediate objective" of "protecting vital military installations."[54] Professor Freedman proceeds to charge the creator and supporters of the Initiative (presumably not with reference to any "intermediate objective") with

> the nurturing of illusions about some technical fix to the problem of vulnerability in the nuclear age.[55]

The range of objectives pertinent to the SDI are, overall, more likely to be achieved than are the objectives typically advertised with reference to the East–West arms control process. However, it is not sensible to condemn that process on the grounds that some people choose to nurture illusions about millennial possibilities that might lurk in Geneva.[56] The SDI is not a quest after the absolute weapon, it does not carry the promise of repealing the nuclear age, and it does not rest on the assumption that solving problems in military technology by military technology somehow, miraculously, translates into solving fundamental problems of political antagonism between states. The pursuit of effective strategic defenses is no more an exercise in escapism than the tactical development of the storm battalion and the technical invention of the tank were efforts to escape the stalemate in *flankless* trench warfare.

The long-range, nuclear-armed ballistic missile is a military–tactical instrument capable, in specified circumstances, of delivering swift victory to an aggressor. More to the point, the unique military characteristics of the ICBM and SLBM, in suitable tactical combination, provide fuel for briefings for political leaders who are seeking to escape the bonds of deterrence. The SDI, as an intermediate objective, most certainly could generate an architecture of weapons and supporting C^3I suitable to frustrate Soviet military (and hence political) objectives in war. Ambitions for offensive counterforce prowess, rendered difficult enough by the developments in hardening, proliferation of facilities, and agility in deployment mentioned already, should be terminally discouraged, even by early deployment of active missile defenses that would be highly cost effective in protecting assets of limited individual value.

It may be objected that active defense of strategic retaliatory forces and C^3I facilities is unnecessary: the superhardening, proliferation, and mobility options already cited should provide convincing material for the discouragement of first-strike adventure. Some of the more important changes in the United States target base for Soviet defense planners ought, as claimed above, to encourage the Soviet Union to place diminishing reliance on a preemptive counterforce blow. But, the benefits of *intermediate* strategic defensive deployments for the United States become self-evident once one takes realistic note of the domestic American political context.

First, although it is true that United States silo-housed ICBMs can be launched very promptly after (confirmation of) attack, as the Scowcroft Report claims,[57] American politicians are not comfortable with a missile force that would have to be used so promptly or lost to the enemy's attack. Strictly speaking, because of C^3I modernization, MX in silos will not be a vulnerable force. Nonetheless, the American political system has been educated over the years to equate survivability with the ability to ride out an attack. Terminal and/or late mid-course active defenses could assure restoration of the now traditional meaning of a survivable ICBM force.

Second, it has yet to be established that the United States public will tolerate mobile ICBM deployment. The new small missile (generically called *Midgetman*) on which Congress thus far has insisted: (1) may be too expensive in an agile basing mode; (2) may meet a fatal level of political opposition for reason of its public interface; and (3) may be unacceptably vulnerable to the threat of Soviet barrage attack. The United States ability to replicate, functionally, the Soviet SS–20, SS–24, and SS–25 has to be judged very much in doubt.

Politics, not technology, is the problem. Strategic defense for the ICBM force probably is technically more reliable, more cost effective, and would be (domestically) politically easier to manage than would a mobility program.

Third, without active defense assistance, the timelines for effective action by the National Command Authorities (NCA) in receipt of warning of attack are very tight but theoretically adequate. Modest levels of defense could buy critically valuable additional minutes for confirmation and assessment of attack, deliberation, and dissemination and execution of launch orders (emergency action messages). If a war is to be conducted according to the near real-time wishes of central political authority—and not mechanistically according to some facsimile of *The (Predetermined) Plan*—it is important that what can be done should be done to ensure reliable continuity of that authority. The Soviet Union has undertaken an extraordinary construction effort to provide superhard and hard bunker protection for its leaders. That option is not politically available to the United States government today.

The United States SDI, contrary to Lawrence Freedman's repeated assertion, has not effected a retreat from the goal of defending people to the intermediate goal, discussed briefly above, of protecting vital military installations. Quite to the contrary, it is understood that it is essential that the SDI should pose an open-ended threat to long-range missiles, whatever their missions. The Western Alliance should not forget that *nuclear* deterrence is but a means to the end of preventing war. Technological change in the balance of advantage between offense and defense, even to the point at which, in President Reagan's words, ballistic missiles are rendered "impotent and obsolete,"[58] need not challenge deterrence itself. The policy aspiration for strategic defense is that the deterrence of war may be achievable possibly more reliably, and certainly by safer means, by a strategic posture dominated by defensive nonnuclear, rather than offensive nuclear, weapons.

The technological promise in strategic defense is that it may, though only may, transform for the better the condition of United States and Allied security. The active defense of population probably is considerably easier to achieve than typically is recognized, if only because public discussion of urban area defense virtually neglects to incorporate genuinely strategic argument. If the Soviet Union should anticipate the strong possibilities: (1) that active defenses will deny it a militarily useful degree of reliable access to United States military targets; (2) that United States urban assets will be heavily defended also; and (3) that United States strategic offensive forces *may* be capable of evading or otherwise defeating Soviet active defenses, what choices would Soviet leaders have? A preferred Soviet choice is not difficult to identify. As in 1972 with SALT I, the Soviet Union should be expected to press for an arms control regime that would minimize their military disadvantage, actual and impending, and so restrict the more leading ominous elements in the threat that time would be bought for Soviet science-based industry to effect a technological "catch up." If United States SDI weaponization cannot totally be precluded, the Soviet Union will endeavor to constrain it to kinds of activity in which Soviet capability at least is competitive; that is, to terminal and late mid-course defense.[59]

A true crisis in policy and strategy would be caused for Moscow if the United States was adamantly unwilling to place any constraints on strategic defensive weapon development, testing, or deployment. Soviet defense planners are not likely to be attracted by George A. Keyworth's idea that strategic defense would relegate ballistic missiles to the status of "a retaliatory deterrent [and] not a preemptive threat."[60] The proposition that strategic defenses would have the effect of compelling a transition in nuclear targeting from military to civilian facilities neglects the culturally rooted *war doctrine* of the Soviet Union.[61] To focus nuclear targeting on American cities, because American military installations were not accessible, would neither contribute usefully to the achievement of military missions, nor would it be an inexpensive enterprise.

Soviet planners would hardly regard as an efficient allocation of scarce nuclear assets the prospect of having to duel with four or more layers of a weaponized United States SDI—with battle space extending out from the boost phase through mid-course to the high endoatmospheric region[62]—in order to ensure warhead penetration to destroy an uncertain fraction of urban America, a target set generally of tertiary or even lesser priority. It is much more probable that the Soviet Union would respond to an emerging United States offense–defense advantage by: (1) redoubling its technological and tactical efforts to maintain penetration survivability for a preemptive counterforce blow; (2) effecting an orderly shift of military reliance from offensive to defensive counterforce (plus passive protection where feasible) in the likely circumstance that (1) was unpromising; and (3) seeking large-scale negotiated reductions in offensive forces in order to negate any United States advantage in the penetration survivability of its offensive weapons. The United States government should be seriously interested in paying a large price in capability to penetrate Soviet defenses, in return for large reductions in Soviet offensive forces.[63]

To achieve rough parity in strategic effectiveness, the Soviet Union would not require defenses technologically comparable to United

States defenses: rather, it would need defenses capable of imposing discouragingly heavy attrition on United States offensive forces. Mutually agreed reductions in the offense should have the effect of reimposing a tolerably even competence in prowess, as between heavily deployed defenses of different degrees of technical sophistication. American political leaders should be expected to be willing to trade a measure of offense–defense advantage for a degree of greater assurance that United States defenses actually could defend people.[64]

The Soviet Union has never eschewed pursuit of a classical strategy that looks to the defeat of the enemy *and* to the protection of critical home values. Even for the United States, the idea of mutual vulnerability of societies as a *desideratum* has been of relatively brief duration (its authoritative sway roughly encompasses the period 1966–82/3). The ABM Treaty of 1972, which proclaims in Article 1 (2) that "each Party undertakes not to deploy ABM systems for a defense of the territory of its country and not to provide a base for such a defense . . ." stands in a condition of some doctrinal tension with the undiminished Soviet programs of deployment for nation-wide air defense, the momentum behind their civil defense preparations, and the scale and scope of their BMD research over the past decade.[65]

It would be an error to equate a condition of offense dominance necessarily with acquiescence in the theory that *mutual* vulnerability is either extant operational reality or is desirable. Soviet war doctrine, as expressed on its leading edge in the quality and quantity of long-range ballistic missiles, seeks to limit damage to the Soviet homeland by seizing the initiative through disruptive–destructive preemptive counterforce strikes. Damage limitation can be effected through offensive and formally defensive weaponry. It would be unwise to endeavor to predict the future of the strategic arms competition in detail, but it is probably reasonably safe to claim that, by the end of the twentieth century, the late 1960s' and 1970s' equation in the United States of nuclear deterrence (and indeed of deterrence more broadly) with the mutual vulnerability of societies will be seen to have been but a fashionable, doctrinal rationalization of a transient phase in the history of military technology.[66] Any readers of this essay who view themselves as seeking heroically to plug holes in the dyke of mutual assured destruction truth against erosion by the purveyors of essentially prenuclear military ideas, should ponder the following judgments of Lawrence Martin:

> A doctrine of mutual deterrence, implemented in this way [threats to cities], requires nations to live for the rest of history with their survival entirely dependent on the good sense and restraint of their

opposite numbers. Such a system seems unlikely to be one conducive to a relaxed international environment. A sudden change in government, an unexpectedly severe crisis, may overturn all expectations. Rich and inventive nations will surely be continually tempted to search for technical or doctrinal ways of reasserting greater control over their fate.[67]

In and of themselves, defense-heavy, as contrasted with offense-exclusive, United States strategic forces would not reflect a revolution in strategic ideas. The authoritative idea at the heart of the countervailing strategy is the proposition that deterrent effect is maximized when a potential enemy anticipates the defeat of his arms.[68] Whether that defeat is imposed by offensive or defensive counterforce, by passive protective measures, or by some mix of all three, is really beside the point. What is not beside the point are the implications of new Western defensive deployments for Soviet decisions on resource allocation between offense and defense. If the Soviet Union chooses, for the most pressing of prudent military reasons, to shift the balance of its effort between offensive and defensive counterforce in favor of the latter—an option that may well be exercised—then the Western Alliance will have to face the fact that the overall terms of deterrence will change. Swords and shields should be fully interchangeable as to effectiveness, if the deterrence problem strictly is to discourage or preclude nuclear offensive action by the other side. But, as this discussion has noted already, the long-range nuclear bombardment forces of the West do not, at present, have such a limited mandate.

Long-range nuclear threats will always lurk as a backstop to defensive deterrence, albeit a backstop that may, indeed should, lack credibility (though the current U.S. offensive-force threat implicit in extended deterrence duties is not exactly a shining example of high plausibility). However, the burden of deterrence, as Soviet strategic defenses improve and thicken, will have to devolve on Western military capabilities that have not been overtaken by technological and tactical innovation. It is true that a weaponized SDI should be able to perform a most useful service in protecting general purpose forces against massive preemptive disruption and damage and thereby should facilitate effective resistance.[69] Nonetheless, the price the United States and its allies would pay for the benefits of a world of strategic (and medium-range theater) defense dominance, would not be a light one. International order, to a degree unprecedented since the end of World War II, would depend most heavily on perceptions, particularly on Soviet perceptions, of the putative defensive prowess of Western conventional and battlefield nuclear weapons.

SOME PROPOSITIONS

The history of military technology in its complex relation to policy suggests nine concluding points that are best described as propositions.[70]

1. *In the words of Raymond L. Garthoff: "Neither side has ever attempted to, and probably could not adequately, determine what the other sees as its deterrence and defense requirements, and it is irresponsible to claim to have done so and thus to be able to pronounce objectively on the requirements of the other side."*[71] This is a considerable overstatement, and its boldness on the Soviet side greatly exceeds the evidence. Nonetheless, Garthoff points usefully to the enduring problem of ethnocentrism in the intelligence input to defense planning and strategic policy more generally.[72] The history of the twentieth century is littered with war plans that failed in execution, to an important degree, because they rested for their operational validity on inexplicit false, or at least highly misleading, cultural stereotypes of enemies and potential enemies. Like defense planners in Berlin and Paris before 1914,[73] those in Washington (and presumably in Moscow) in the 1980s are prone to predict adversary behavior that is convenient for the prospects of success for *our* arms.

2. *The technical details of calculated advantage and disadvantage between armed forces cannot serve as reliable guides to the likelihood of war.* In other words, the typically very narrowly defined concept of strategic stability, so beloved by American defense and arms control analysts of most shades of doctrinal preference, should not be regarded as *the* key to understanding the political meaning of weaponry. There are several problems with the concept of strategic stability as an objective attribute of some, rather than other, military capabilities.

First and foremost, what matters for decisions on peace or war is what policy makers believe, not what only the passage of time will or may reveal actually to be the case. The military technology of 1914, in conjunction with mass armies and alliance structures (meaning that an important ally facing defeat could expect assistance to keep it "in the field"), denied the armies of the Great Powers the status of being instruments of rapid decision. Unfortunately, the general staffs and the governments they influenced either did not know that, or chose not to know it, ahead of the trial by battle.[74]

Second, although it may be critically important for war or peace whether statesmen believe that their military instruments can solve political problems at bearable national cost: (1) countries do not go to war *because* their general staffs predict military success (the plausibility of delivery of a real-world referent for a suitable definition of victory may be a necessary, but it is not a sufficient, cause of war);

(2) the strategic balance relevant to major policy decisions on war and peace is likely to encompass a much wider range of concerns than, for a *reductio ad absurdum*, the balance in surviving equivalent megatonnage following a hypothetical first strike by either side.

Third, stability analysis does not yield reliable transcultural *truths*. This third problem is well illustrated in the following gem from a recent article in *Pravda*:

> Now the Pentagon plans to resurrect "deterrence [through terror]" by protecting U.S. missiles with space weapons. And no matter how much the proponents of "star wars" insist that these weapons are defensive, it is impossible to believe it. Reducing the vulnerability of strategic weapons whose mission it is to destroy the opponents launchers is tantamount to destroying the parity which now exists, to creating a new potential for a first strike. This is precisely what the ABM Treaty was designed to prevent, in order to deter nuclear aggression.[75]

3. *The ability of military intelligence staffs to predict how weapons would be used in operational plans in the future is unlikely to be superior to the record of prewar intelligence assessments in the twentieth century thus far.* Campaign strategy is more likely to provide cause for surprise than are the details of performance of individual pieces of military equipment. Weapons, to be effective for deterrence, should carry the plausible prospect of defeating, or helping to defeat, the enemy's strategy on his terms. A great deal of what passes for strategic analysis in the United States is not worthy of the name, because it neglects political factors and even much of military operational reality. The persisting debate over ICBM vulnerability is a classic case in point.[76]

4. *Understanding of which kinds of threat capability and behavior are most likely to deter (whom, from doing what?) remains profoundly rudimentary.*[77] Inputs to policy on threat behavior frequently are unsystematically anecdotal in character, rest heavily on possibly irrelevant historical analogies (that may not even be recalled accurately),[78] are colored by cultural stereotyping, and can be highly personal and intuitive. This is not to imply that any top policy-making group necessarily can wage crises in a far more "rational" manner than is alleged here to be typical. But, it is to suggest that the role of particular weapons, and of particular military threats in deterrence, may be a great deal more variable than the strategic policy debate could lead one, unwisely, to believe.

5. *The pendulum of relative advantage between the strategic offense and the strategic defense is likely to swing back toward the latter over the next several decades.* Given the extremity of offense dominance in the ballistic missile realm through the 1960s, 1970s, and to date in the

1980s, it is fair to say that the metaphorical pendulum at issue cannot swing in any other direction.[79] One may suspect that the defense is very likely to achieve high competence, if only because it has near total technical growth potential as compared with the far more mature technologies of the offense. However, technological history does not proceed in orderly cycles from dominance by offense to dominance by defense and then back again. Strategic defense may achieve high competence for some relatively undemanding military missions, yet only a very debatable measure of effectiveness for more difficult tasks. Furthermore, the technical requirements and even the implications for a policy of defense dominance have to be identified specifically for a nuclear environment.

6. *"Confidence in conventional deterrence by defense is unwarranted because denial is inherently a weaker deterrent than punishment—the costs of failure are much smaller."*[80] This judgment by Richard Betts is quoted as a proposition worth discussing, not in any sense as a conclusion. On balance, this author finds the proposition more interesting than plausible. Denial strategies may lack a close audit trail to society-wide catastrophe, and hence deterrence may be said to be weaker than in the case of contingent promises of (nuclear) punishment, but they are inherently credible as to prospective use. If, as seems most likely, long-range nuclear threats—whatever the delivery vehicles may be for nuclear weapons—become decreasingly credible as active defenses achieve greater competence, the Western Alliance will have no prudent choice other than to seek more capable conventional forces. Of course, residual nuclear dangers as a source of deterrent effect are bound to remain, virtually regardless of the quantity and supposed quality of strategic defenses. There will be anxiety over the possibility of innovations in tactics, and hard-to-reduce suspicions will endure to the effect that Clausewitz's "friction" factors might serve to render actual defensive system performance dramatically less effective than expected (and required). Nonetheless, a fairly steady drift in competitive advantage toward the strategic defense could place the West in a condition in which the consequences of a failure of deterrence could, and even should, be far less severe than is the case today, but in which the probability of deterrence failure might be correspondingly higher. This, of course, may lead one into the problem area discussed under proposition 2.

War will not occur simply because the rival alliances change the military terms of deterrence. But, in a political situation in which Soviet leaders are genuinely interested in exploring their military options, a framework of deterrence that is far more heavily dependent than today's on (Soviet) beliefs concerning the combat effectiveness of NATO's conventional forces could be a framework of deterrence that would be difficult to maintain. In his extensive historical study of conventional deterrence, John Mearsheimer judges that "in a crisis, if one side has the capability to launch a blitzkrieg, deterrence is likely to fail."[81] But, on the brighter side, "deterrence is likely to hold when a potential attacker is faced with the prospect of employing an attrition strategy, largely because of the associated exorbitant costs and because of the difficulty of accurately predicting ultimate success in a protracted war."[82] Nonetheless, optimism should be bounded: "when a nation has a powerful motive for war, it will go to great lengths to find a suitable military strategy,"[83] and, one might add, such a nation is more likely than not to persuade itself that it has a suitable strategy, whether or not it does (remember the example of Japanese "best case" analysis in 1941).[84] It is worth noting that the more capably strategic defenses could protect Western societies, the more capably they could protect the defense mobilization base of the Western maritime alliance for the conduct of a protracted war.

7. *It may well be the human lot to have to live prospectively forever with nuclear weapons, but the terms of the relationship with nuclear dangers should not be beyond a useful measure of control.* In company with political endeavors to reduce incentives to fight and to provide more sensible "rules of the road" for the state behavior in matters of international security, there is everything to be said in favor of seeking whatever technological safety equipment can be devised. There should be no danger that a process of transition, prudently effected, to a defense-heavy strategic posture would create new dangers of war occurring. But, as outlined tersely in proposition 6, there could be some new quality of strain for conventionally oriented, though not exclusive, deterrence around peripheral Eurasia. If that anxiety is sustained by the conclusions of further study, though one hopes not by real-world experience, it has to be considered in the context both of the benefits of the strategic defensive transition and of the extensive (several decades) period of time available to NATO to place its conventional (and short-range nuclear and chemical) house in much better order.

There is no necessity for resignation to an alleged certainty, or effective certainty, either that a Soviet–American war must be a nuclear war, or that a nuclear war must be a holocaust. Clausewitz observed:

> War is an act of force, and there is no *logical* limit to the application of that force. Each side, therefore, compels its opponent to follow suit; a reciprocal action is started which must lead, in *theory*, to extremes. (Emphasis added.)[85]

A little earlier, Clausewitz explained:

> If, then, civilized nations do not put their prisoners to death or devastate cities and countries, it is because intelligence plays a larger part in their methods of warfare and has taught them more effective ways of using force than the crude expression of instinct.[86]

In addition to self-regarding restraint in the realm of the application of nuclear force and to considerations of military efficiency in that application, the technology of nuclear weapon delivery may well prove to be vulnerable, indeed massively and perhaps comprehensibly vulnerable, to the weapons and tactics of nonnuclear strategic defense.

8. *Progress in an arms control process will not be achieved via a quest for strategic "truth"; be that truth allegedly of an enduring or a transitory character. Instead, progress in an arms control process will be fueled by calculations of net military and political advantage (and of net disadvantage temporarily reduced or precluded).* Anticipation of a military disadvantage tomorrow (the disadvantages of an emerging strategic context may be mutual, depending on operational circumstances), which can be addressed usefully today, is the engine of arms control success. In and of itself, the control of military technology is not a goal of arms control. Just as the audit trail is faint from the details of strategic balance to decisions for war or peace, so the belief that "progress" in an arms control process is important for war or peace is a difficult one to sustain under close scrutiny. Cause and effect can be confused. Frequently, states that discern a more than trivial prospect of war occurring between them compete in armaments. Similarly, states that wish to express visibly a strong desire to improve their political relations frequently find it convenient to negotiate arms control agreements.

9. *Rapid military–technological change is not, inherently, a threat to peace or even to strategic stability (narrowly technically defined).* One can always find candidates for exceptions to the proposition that technological change is benign (which is the reason for the inclusion above of the qualifier *inherently*). Generically, military postures in permanent flux with new weaponry, and in anticipation of new weaponry, are military postures that show far greater promise of success for tomorrow than for today. The potential problem of unevenness in military–technological (*inter alia*) developments between rival powers—the problem that can encourage now-or-never, window-of-opportunity impulses—is not a valid ground for opposing new military technologies. Instead, it suggests that during a period of transition from older to new or newer methods and means of war, great care must be taken to ensure provision of adequate deterrence

cover for the years of change (given the possibility that putative enemies may need more deterring than before).

NOTES

1. The discussion of the relationship between strategic offense and defense necessarily occupies considerable space in the text. Although this chapter is analysis and not advocacy, the author has not sought to conceal his belief that President Reagan's Strategic Defense Initiative (SDI) is a strategic policy necessity. See Colin S. Gray, "Strategic Defense, Deterrence, and the Prospects for Peace," *Ethics*, vol. 95, no. 3 (March 1985), pp. 659–72.

2. McGeorge Bundy, "Existential Deterrence and Its Consequences," in *The Security Gamble: Deterrence Dilemmas in the Nuclear Age*, ed. Douglas MacLean (Totowa, N.J.: Rowman and Allanheld, 1984), pp. 3–13.

3. This hypothesis is argued strongly in Stephen Peter Rosen, "Foreign Policy and Nuclear Weapons: The Case for Strategic Defenses," in *The Strategic Imperative: New Policies for American Security*, ed. Samuel P. Huntington (Cambridge, Mass.: Ballinger, 1982), pp. 141–61.

4. For example, see H. W. Koch, ed., *The Origins of the First World War: Great Power Rivalry and German War Aims*, 2d ed. (London: Macmillan, 1984); L. L. Farrar, Jr., *Arrogance and Anxiety: The Ambivalence of German Power, 1848–1914* (Iowa City: University of Iowa Press, 1981); and Williamson Murray, *The Change in the European Balance of Power, 1938–1939: The Path to Ruin* (Princeton, N.J.: Princeton University Press, 1984).

5. See Michael Howard, "Men against Fire: Expectations of War in 1914," *International Security*, vol. 9, no. 1 (Summer 1984), pp. 41–57; and Colin S. Gray, *Protracted War: The Lessons of History* (Fairfax, Va.: National Institute for Public Policy, 1984).

6. See Jack L. Snyder, *The Ideology of the Offensive: Military Decision Making and the Disasters of 1914* (Ithaca, N.Y.: Cornell University Press, 1984).

7. Compare and contrast Bernard Brodie, ed., *The Absolute Weapons: Atomic Power and World Order* (New York: Harcourt Brace, 1946), and William Liscum Borden, *There Will Be No Time: The Revolution in Strategy* (New York: Macmillan, 1946). For analytical overviews of the history of nuclear strategy, see Lawrence Freedman, *The Evolution of Nuclear Strategy* (London: Macmillan, 1981); and Fred Kaplan, *The Wizards of Armageddon* (New York: Simon and Schuster, 1983).

8. Particularly useful is John Terraine, *White Heat: The New Warfare, 1914–18* (London: Sidgwick and Jackson, 1982).

9. Bundy, "Existential Deterrence and Its Consequences." Freedman's position is not that different from Bundy's: *The Evolution of Nuclear Strategy*, p. 400.

10. A charge leveled in Robert Jervis, *The Illogic of American Nuclear Strategy* (Ithaca, N.Y.: Cornell University Press, 1984), pp. 56–63.

11. Ibid., p. 56.

12. Note the cautionary analysis in Robert Gilpin, *War and Change in World Politics* (Cambridge: Cambridge University Press, 1981), particularly p. 217. Gilpin writes: "We are but a few decades into the nuclear age, and it is far too early to conclude that there will not be a Gaius Marius, Alexander, or Napoleon who will develop tactics and strategy to make nuclear weapons and the nuclear threat effective instruments of national policy."

13. On the concept of "fighting power" see Martin van Creveld, *Fighting Power: German and U.S. Army Performance, 1939–1945* (Westport, Conn.: Greenwood, 1982).

14. Carl von Clausewitz, *On War*, ed. Michael Howard and Peter Paret (Princeton, N.J.: Princeton University Press, 1976), pp. 119–21.

15. Judicious recognition of the uncertainties permeates The National Academy of Sciences, *The Effects on the Atmosphere of a Major Nuclear Exchange* (Washington, D.C.: National Academy Press, 1985); and U.S. Department of Defense, *The Potential Effects of Nuclear War on the Climate* (Washington, D.C.: Department of Defense, March 1985). Scientific uncertainties are given more abrupt treatment in Paul Ehrlich et al., *The Cold and the Dark: The World after Nuclear War* (New York: Norton, 1984). A devastating brief critique of Ehrlich et al. is the review by George W. Rathjens and Ronald H. Siegel in *Survival*, vol. XXVII, no. 1 (January–February 1984), pp. 43–4.

16. See Colin S. Gray, "The Nuclear Winter Thesis and U.S. Strategic Policy," *The Washington Quarterly*, vol. 8, no. 3 (Summer 1985), pp. 85–96.

17. Terraine, *White Heat*, p. 93.

18. The judgment encompasses the tanks of the day, the artillery tactics perfected by both sides (by 1917–18), and the "new" infantry tactic of infiltration (developed and applied by the Germans at Cambrai in November–December 1917 and in their great spring offensive in 1918).

19. This author recognizes fully that it is an open question whether or not nuclear threat and effectively total societal vulnerability can usefully be treated as a problem rather than a condition. The condition hypothesis permeates The Harvard Nuclear Study Group, *Living with Nuclear Weapons* (New York: Bantam, 1983); and, for a pithy treatment, Christoph Bertram, "Strategic Escapism," *The Washington Post*, March 17, 1985, p. E8.

20. See Philippe Contamine, *War in the Middle Ages* (Oxford: Basil Blackwell, 1984; first pub. 1980), chapter 6.

21. See Georg Bruchmüller, *The German Artillery in the Breakthrough Battles of the World War* (J. H. Wallace and H. D. Kern trans. of 2nd [1922] ed.) (Ft. Sill, Okla.: U.S. Army Field Artillery School, n.d.). The most effective artillery barrage in history was conducted by Bruchmüller in two and a half hours on May 27, 1918, on the Aisne.

22. See George H. Quester, *Deterrence before Hiroshima: The Airpower Background of Modern Strategy* (New York: John Wiley, 1966).

23. The railroad, machine tools to make the machines for assembly-line mass production, the internal combustion engine, the computer, and so on.

24. The German High Command had to learn from very expensive experience in 1941 that the breakthrough potential of *panzer* forces on the

Eastern Front had an entirely different strategic meaning from that potential as developed in practice in Poland and France in 1939 and 1940. See Albert Seaton, *The Russo–German War, 1941–45* (New York: Praeger, 1970), pp. 190, 202.

25. On the trend toward a *classical strategy* of offense–defense see Colin S. Gray, "War-Fighting for Deterrence," *The Journal of Strategic Studies*, vol. 7, no. 1 (March 1984), pp. 5–28.

26. See Spurgeon M. Keeny, Jr., and Wolfgang K. H. Panofsky, "MAD Versus NUTS," *Foreign Affairs*, vol. 60, no. 2 (Winter 1981/82), pp. 287–304.

27. In his Report for FY 1986, Secretary of Defense Weinberger has written as follows: "Of the three military sources of deterrence, the most reassuring is effective defense [the others are "the threat of escalation" and "the threat of retaliation"]. A potential aggressor would have little cause to doubt that a nation would use its military capabilities to try to repel an attack. Effective defense is also less likely to cause escalation than are the other two sources of deterrence, and it provides the means for protecting ourselves should deterrence fail." *Annual Report to the Congress, Fiscal Year 1986* (Washington, D.C.: USGPO, February 4, 1985), p. 26.

28. Michael Howard has observed as follows: "Society may have accepted killing as a legitimate instrument of state policy, but not, as yet, suicide. For that reason I find it hard to believe that the abolition of nuclear weapons, even if it were possible, would be an unmixed blessing. Nothing that makes it easier for statesmen to regard war as a feasible instrument of state policy, one from which they stand to gain rather than lose, is likely to contribute to a lasting peace." *The Causes of Wars and Other Essays* (London: Counterpoint, 1984; first pub. 1983), p. 22.

29. See Jervis's discussion of "Why Deterrence Is Not So Difficult," in *The Illogic of American Nuclear Strategy*, pp. 153–57.

30. Useful commentaries include: J. Michael Legge, *Theater Nuclear Weapon and the NATO Strategy of Flexible Response*, R–2964–FF (Santa Monica, Calif.: RAND, April 1983); John D. Steinbruner and Leon V. Sigal, ed., *Alliance Security: NATO and the No-First-Use Question* (Washington, D.C.: Brookings, 1983); and David N. Schwartz, *NATO's Nuclear Dilemmas* (Washington, D.C.: Brookings, 1983). I have challenged the utility and prudence of a heavily nuclear-dependent flexible response in my forthcoming book, *Geopolitics, Strategic Culture, and National Security*.

31. See Walter Slocombe, "Extended Deterrence," *The Washington Quarterly*, vol. 7, no. 4 (Fall 1984), pp. 93–103.

32. Albert Wohlstetter, "The Delicate Balance of Terror," *Foreign Affairs*, vol. 37, no. 2 (January 1959), pp. 211–34.

33. For a thoughtful historical analysis of this issue, see Richard Ned Lebow, "Windows of Opportunity: Do States Jump through Them?" *International Security*, vol. 9, no. 1 (Summer 1984), pp. 147–88.

34. See Vernon V. Aspaturian, "The Anatomy of the Soviet Empire: Vulnerabilities and Strengths," in *Military Strategy in Transition: Defense and Deterrence in the 1980s*, ed. Keith A. Dunn and William O. Staudenmaier

(Carlisle Barracks, Pa.: U.S. Army War College, 1984), pp. 97–146; and Richard Pipes, *Survival Is Not Enough: Soviet Realities and America's Future* (New York: Simon and Schuster, 1984).

35. Howard, *The Causes of Wars*, particularly pp. 142–43.

36. The controversial articles in question are: "Nuclear Strategy: The Case for a Theory of Victory," *International Security*, vol. 4, no. 1 (Summer 1979), pp. 54–87; and, with Keith Payne, "Victory Is Possible," *Foreign Policy* 39 (Summer 1980), pp. 14–27.

37. Harry G. Summers, Jr., *On Strategy: A Critical Analysis of the Vietnam War* (Novato, Calif.: Presidio, 1982), pp. 63–9.

38. *Annual Report to the Congress, Fiscal Year 1986*, p. 27.

39. Ibid. Weinberger is more blunt than this: " 'Favorable' means that if war is forced upon us, we must win—we cannot allow aggression to benefit the aggressor. It does not mean more territory or power for the United States."

40. Thomas Schelling, *Arms and Influence* (New Haven, Conn.: Yale University Press, 1966), pp. 69–91.

41. See Samuel P. Huntington, *Instability at the Non-Strategic Level of Conflict* (Washington, D.C.: Special Studies Group, Institute for Defense Analyses, October 6, 1961). If the United States and its allies effectively are inferior in land forces, are unable to translate command of the sea into military advantage on land, and enjoy no more than parity in strategic forces, the overall balance cannot be said to be satisfactory.

42. A useful treatment is Edward N. Luttwak, *The Pentagon and the Art of War* (New York: Simon and Schuster, 1984), pp. 229–36.

43. Herman Kahn, *On Escalation: Metaphors and Scenarios* (New York: Praeger, 1965), pp. 23, 24, 290.

44. See Thomas Powers, "Choosing a Strategy for World War III," *The Atlantic*, November 1982, pp. 82–110; Jeffrey Richelson, "PD–59, NSDD–13 and the Reagan Strategic Modernization Program," *The Journal of Strategic Studies*, vol. 6, no. 2 (June 1983), pp. 125–46; Peter Pringle and William Arkin, *SIOP* (London: Sphere, 1983); Desmond Ball, *Targeting for Nuclear Deterrence*, Adelphi Papers No. 185 (London: IISS, Summer 1983); and Leon Sloss and Marc Dean Millot, "U.S. Nuclear Strategy in Evolution," *Strategic Review*, vol. XII, no. 1 (Winter 1984), pp. 19–28; and Colin S. Gray, "Strategic Forces," in *American Defense Annual, 1986–1987*, ed. Joseph Kruzel (Lexington, Mass.: Lexington Books, 1986).

45. The President's Commission on Strategic Forces, *Report* (Washington, D.C.: The White House, April 1983), p. 7.

46. Ibid., p. 6.

47. A thesis advanced and explained in: Blair Stewart, "Peacekeeper," Defense Systems *Review and Military Communications*, June 1984, pp. 46–52; and Colin S. Gray, *Missiles Against War: The ICBM Debate Today* (Fairfax, Va.: National Institute for Public Policy, 1985).

48. For details of these and other Soviet strategic offensive systems, see Caspar W. Weinberger, *Soviet Military Power, 1985* (Washington, D.C.: USGPO, April 1985), chapter 2.

49. See Admiral James D. Watkins, "The Maritime Strategy," in

Watkins et al., *The Maritime Strategy* Supplement to *U.S. Naval Institute Proceedings* (January 1986), pp. 11, 14.

50. Robert Jervis, "MAD is the best possible deterrence," *Bulletin of the Atomic Scientists*, vol. 41, no. 3 (March 1985), pp. 43–5.

51. See Michael F. Altfeld and Stephen J. Cimbala, "Targeting for Nuclear Winter: A Speculative Essay," *Parameters*, vol. XV, no. 3 (Autumn 1985), pp. 8–15.

52. See Michael J. Deane, *Strategic Defense in Soviet Strategy* (Washington, D.C.: Advanced International Study Institute [in association with the University of Miami], 1980); Sayre Stevens, "The Soviet BMD Program," in *Ballistic Missile Defense*, ed. Ashton B. Carter and David N. Schwartz (Washington, D.C.: Brookings, 1984), pp. 182–220; and Weinberger, *Soviet Military Power, 1985*, chapter 3; and U.S. Department of Defense and Department of State, *Soviet Strategic Defense Programs* (Washington, D.C.: USGPO, October 1985).

53. See Nathan Leites, *Soviet Style in War* (New York: Crane, Russak, 1982); and Colin S. Gray, *Nuclear Strategy and National Style* (Lanham, Md.: Hamilton Press, 1986), chapter 3.

54. Lawrence Freedman, "NATO and the Strategic Defense Initiative," *NATO's Sixteen Nations*, vol. 29, no. 6 (November 1984), pp. 6–7. Also see Professor Freedman's letter to the editor, *The Times* (London), September 21, 1984, p. 11. On March 29, 1985, President Reagan said of the SDI that it "is not, and should never be misconstrued as, just another method of protecting missile silos." Quoted in David Hoffman, "Reagan Seeks to Link SDI, Missile Cutbacks," *The Washington Post*, March 30, 1985, p. A13. Secretary of Defense Caspar Weinberger has written: "The defense that might evolve from the research programs will not be intended to defend our strategic weapons systems. We are considering ways to defend both our territory and that of our allies against the ballistic missile threat." *Annual Report to the Congress, Fiscal Year 1987* (Washington, D.C.: USGPO, February 5, 1986), p. 287.

55. Freedman, "NATO and the Strategic Defense Initiative," p. 7.

56. A subject on which Lawrence Freedman is very cogent and persuasive indeed. See his article, "Arms Control: No Hiding Place," *SAIS Review*, vol. 3, no. 1 (Winter-Spring 1983), pp. 3–11.

57. The President's Commission on Strategic Forces, *Report*, p. 8.

58. See President's Foreword to *The President's Strategic Defense Initiative* (Washington, D.C.: USGPO, January 1985), p. ii.

59. It is noticeable that Soviet criticism of the United States SDI, with particular reference to the Geneva process, has spoken almost exclusively of "space strike weapons" and not of strategic defense per se. The author does not assume, in the text, that the United States will be better able to penetrate Soviet defenses than the Soviet Union will United States defenses: only the possibility is raised here as a healthy anxiety in Soviet minds.

60. George A. Keyworth, "Ganging Up on Star Wars," *The Washington Post*, December 24, 1984, p. 15.

61. I have borrowed "war doctrine" from John Erickson, "Threat Identification and Strategic Appraisal by the Soviet Union, 1930–1941," in *Knowing One's Enemies: Intelligence Assessment before the Two World Wars*, ed.

Ernest R. May (Princeton, N.J.: Princeton University Press, 1984), pp. 375–423.

62. Critics of the SDI assert frequently that the United States is exploring the technical feasibility of deploying a Maginot line in space. This popular charge is both incorrect as analogy and grossly unfair to the historical referent. A full-scale weaponized SDI would be a Maginot line with close to 8,000 miles of battle space for defense in depth against ICBMs.

63. On the problems of a "defensive transition," see Keith B. Payne and Colin S. Gray, "Nuclear Policy and the Defensive Transition," *Foreign Affairs*, vol. 62, no. 4 (Spring 1984), pp. 820–42; and Weinberger, *Annual Report to the Congress, Fiscal Year 1986*, p. 55; Charles L. Glaser, "The Transition to Highly Effective Strategic Defenses." Paper prepared for the Wilson Center conference, "Strategic Defense and Soviet–American Relations," March 10–11, 1986; and Colin S. Gray, "The Transition from Offense to Defense," *The Washington Quarterly*, vol. 9, no. 3 (Summer 1986).

64. The Reagan administration is not being unreasonable when it says that states cannot negotiate about research programs. For several, and probably many, years to come, U.S. negotiators cannot know what the SDI is "worth," and Soviet negotiators cannot know what price they should be willing to pay to arrest or rechannel its evolution.

65. Weinberger, *Soviet Military Power, 1985*, chapter 3.

66. The *apogée* of this doctrinal rationalization was well represented in: John Newhouse, *Cold Dawn: The Story of SALT* (New York: Holt, Rinehart and Winston, 1973); and Jerome H. Kahan, *Security in the Nuclear Age: Developing U.S. Strategic Arms Policy* (Washington, D.C.: Brookings, 1975).

67. Lawrence Martin, *The Two Edged Sword: Armed Force in the Modern World* (London: Weidenfeld and Nicolson, 1982), pp. 18–19.

68. See Walter Slocombe, "The Countervailing Strategy," *International Security*, vol. 5, no. 4 (Spring 1981), pp. 18–27.

69. See Weinberger, *Annual Report to the Congress, Fiscal Year 1986*, p. 56.

70. Rather than summarize the discussion in the essay, the author wishes to use this final section to present nine propositions that help expand the analysis presented thus far and that may serve to spark further reflection and enquiry. Some readers of an earlier version of the essay determinedly misread these propositions as conclusions.

71. Raymond L. Garthoff, *Perspectives on the Strategic Balance*, Staff Paper (Washington, D.C.: Brookings, 1983), p. 34.

72. On this subject see Ken Booth, *Strategy and Ethnocentrism* (London: Croom, Helm, 1979); and May, ed., *Knowing One's Enemies*; and Roman Kolkowicz and Ellen Propper Mickiewicz, ed., *The Soviet Calculus of Nuclear War* (Lexington, Mass.: Lexington Books, 1986).

73. "The great Schlieffen never really took the Belgian army seriously. After all, they were just Belgians. Why should they fight?" J. Steinberg, "A German Plan for the Invasion of Holland and Belgium, 1897," in *The War Plans of the Great Powers, 1880–1914*, ed. Paul M. Kennedy (London: Allen and Unwin, 1979), pp. 159–60.

74. See Snyder, *The Ideology of the Offensive.*

75. S. Men'shikov, " 'Kosmicheskie igry' voenoyo kompleksa [The Military Complex's 'Space Games']," *Pravda*, February 18, 1985, p. 6.

76. The vulnerability of the MX ICBM in silos is predicated on a United States president choosing to permit his missile force to attempt to ride out an attack. This vulnerability thesis loses much of its persuasive edge if one asks: (1) what has been happening in the world (is this a "bolt out of the blue"?); (2) what tactics would the Soviet Union employ in order to attack the United States ICBM force; and (3) why would Soviet leaders not be impressed by United States technical capability to launch its ICBMs on first confirmation of nuclear attack?

77. See Paul Huth and Bruce Russett, "What Makes Deterrence Work? Cases from 1900 to 1980," *World Politics*, vol. XXXVI, no. 4 (July 1984), pp. 496–526.

78. See Richard E. Neustadt and Ernest R. May, *Thinking in Time: The Uses of History for Decision-Makers* (New York: Free Press, 1986).

79. For an historical analysis of the offense–defense relationship, see George H. Quester, *Offense and Defense in the International System* (New York: John Wiley, 1977). Also useful are Trevor N. Dupuy, *The Evolution of Weapons and Warfare* (Indianapolis, Ind.: Bobbs-Merrill, 1980); and William H. McNeill, *The Pursuit of Power: Technology, Armed Force, and Society since A.D. 1000* (Chicago, Ill.: University of Chicago Press, 1982).

80. Richard K. Betts, "Conventional Deterrence: Predictive Uncertainty and Policy Confidence," *World Politics*, vol. XXXVII, no. 2 (January 1985), p. 177.

81. John J. Mearsheimer, *Conventional Deterrence* (Ithaca, N.Y.: Cornell University Press, 1983), p. 203.

82. Ibid., p. 206.

83. Ibid., p. 212.

84. Michael A. Barnhart, "Japanese Intelligence before the Second World War: 'Best Case' Analysis," in *Knowing One's Enemies*, ed. May, pp. 424–55.

85. Clausewitz, *On War*, p. 77.

86. Ibid., p. 76.

CHAPTER 6

Strategy as Politics, Politics as Strategy; Domestic Debates, Statecraft, and Star Wars

ARTHUR A. STEIN

INTRODUCTION

On March 23, 1983, in an address on the defense budget request then before Congress, President Reagan proved again that he "is at once an optimist, a radical, an innovator and a plunger" by describing his new "vision of the future which offers hope."[1] Characterizing reliance "on the specter of retaliation" for preventing nuclear war as "a sad commentary on the human condition," he asked if it would not "be better to save lives than to avenge them?" And he wondered, too, if free people might not be more "secure in the knowledge that their security did not rest upon the threat of instant U.S. retaliation to deter a Soviet attack, that we could intercept and destroy strategic ballistic missiles before they reached our own soil or that of our allies?"

But Reagan's speech represented more than his vision of the future, for it was also a direct response to his administration's immediate political past. Following his landslide reelection victory in 1980, the president found his administration's national security policy under assault. His plans for the modernization of America's strategic

I would like to thank Roman Kolkowicz, Benjamin Lambeth, Joseph Nye, and other conference participants, as well as Amy Davis and Lori Gronich for their thoughts on an earlier draft of this paper. I would also like to acknowledge funding provided by the Committee on Research of the Academic Senate of the Los Angeles Division of the University of California and by the supporters of the conference.

arsenal foundered. The requirements of deterrence and defense had become the subject of new debate. A grass-roots movement for a freeze in weapons deployments swept the country, and critics were attacking America's nuclear strategy as immoral. America's European allies, Congress, and ordinary citizens all pressed the president to change his approach to national security and arms control.

In one speech, President Reagan transformed the domestic political debate about America's strategic doctrine and national security policy. He recognized that his goal of shielding the United States from attack might "not be accomplished before the end of this century" and was not without significant political and technical problems. But he accomplished a much more immediate objective. By calling on "the scientific community . . . those who gave us nuclear weapons, to turn their great talents now to the cause of mankind and world peace, to give us the means of rendering these nuclear weapons impotent and obsolete," he changed the political agenda.[2]

Since March 1983, scholars, strategists, politicians, and others have argued about the wisdom of developing defensive space weapons. They have attacked or advocated the president's plan—called Star Wars by its critics and the Strategic Defense Initiative (SDI) by the administration—rather than continue to focus on the issues that concerned them earlier. The desirability, feasibility, viability, practicality, affordability, and optimality of Reagan's new policy have all been considered. His proposal has shifted the focus of domestic political debate and scrambled political positions and coalitions.

Indeed, Reagan's speech created a national debate. For "the Reagan of star wars and supply side is a Reagan at odds with the intellectual poobahs of defense and economics. He is, instead, a president willing to identify with fringe thinkers of both disciplines who are generally classified as rightwing."[3] Indeed, critics immediately dubbed the plan Star Wars and called it crazy. The president subsequently named his proposal the Strategic Defense Initiative (SDI), and supporters now trumpet it as a peace shield. Given the Soviets' evident concern about SDI, critics recommend that the president use it as a bargaining chip and negotiate it away in exchange for reductions in Soviet offensive weapons. Supporters insist that this new vision, of a future without nuclear weapons, is our only hope and must not be dealt away. Negotiators, they argue, should focus on shifting the superpowers toward a world emphasizing defense, rather than continue to stress the mutual assured destruction (MAD) of a world armed almost entirely by offensive strategic nuclear weapons.

This chapter emphasizes that the debate over Star Wars is a political one. To understand and explain the Reagan administration's advocacy of the Strategic Defense Initiative requires an understanding of the relationship between politics and strategy and, most specifically, of the political problems confronting the Reagan administration during its first two years in office. The reason for the president's address was political, and he succeeded brilliantly in shifting the national debate away from the MX and the nuclear freeze and onto the problem of the future. In doing so, he changed his image from that of an unreconstructed warmonger to that of an unreconstructed peacemonger. In addition, the debate over strategic doctrine demonstrates the absence of a consensus among defense analysts, and this implies that national policy will be made (or chosen) on political rather than technical grounds.[4]

STRATEGY AND POLITICS

Issues of national security, especially nuclear strategy, are often treated as technical problems for which appropriate technical solutions exist. Thus, strategists often consider the requisites of defense or deterrence as if these were devoid of political calculations or concerns. To develop a national security policy, however, is to reach an optimal *political* solution to a technical problem. In this way, national security policy is very much like economic policy: politics explains selections of appropriate ends and means more than does science.

The role of science, after all, is limited in policy making. It cannot address the question of which ends societies should pursue. Scientists may have individual opinions about appropriate societal goals, but their objectives do not derive from any scientific data or analysis.[5] Indeed, in American society, their preferences generally carry less weight than those of television and movie stars. Scientists can, however, play a part in guiding a society's choice of the best means for achieving its chosen ends. They can assess whether particular means will achieve given ends and whether the use of those means will also entail other, unintended, consequences.[6] But most typically, either no scientific consensus exists, or there are viable alternatives from which to choose. So, just as they are rarely responsible for the selection of societal objectives, scientists rarely choose the policies intended to ensure those goals. Thus, the choice of social ends and means falls to politicians.

Even when there is a scientific consensus, politicians may choose to ignore it. The national interest, and therefore public policy, encompasses an array of social values and concerns. Scientists, for example, may agree on the contingent truthfulness of the theory of evolution, but religious believers can still mobilize support for having public school teaching reflect their views of creation. And if the mobilization is strong enough, politicians may bow to views they find scientifically laughable. Similarly, public health policy often reflects moral considerations as well as medical ones.[7]

Moreover, politicians choose policies with an eye not only to broad national objectives but also in keeping with individual and group interests. Party concerns and individual careers matter. Even bureaucrats are presumed to consider individual and agency interests in making their assessments. The particularistic interests of the military services, for example, are widely recognized as creating problems in the search for a coherent national security policy.[8] Thus, a politician's *optimal* policy likely optimizes a mix of national, party, and personal objectives. Economic policies, for example, may be chosen not only in order to maximize general economic well-being but also to improve the chances for the reelection of their proponents.[9]

Moreover, policies emerge from compromise. Individual politicians must make choices when tradeoffs exist between objectives. They may be forced to choose, for example, between preventing war and preventing unfavorable changes in the international status quo, or between guns and butter. Not only can national goals conflict, but there also may be tradeoffs between national and particularistic interests. Finally, having made individual choices, politicians must often negotiate and compromise with one another. Thus, politicians rarely begin by proposing the actual policies they want instituted. They start with positions more extreme than their actual desires and establish bargaining positions. Or they attempt to institute one program in order to force positions on another. By securing the 1981 tax cut, for example, the Reagan administration positioned itself perfectly on economic policy and forced the Democrats to choose between cutting domestic spending, increasing taxes, or accepting big current deficits that would endanger future domestic spending. Finally, given all the compromises, it is not surprising that the resulting policies are odd mixtures, concatenations of inconsistent elements.[10]

Similarly, the provision of national defense and security are inherently political questions. Although strategists do emphasize the importance of political objectives when analyzing how war might break out, they do not discuss the political objectives inherent in the adoption or rejection of national security policies. Yet the level of military spending, of how much is enough, has been an annual focus of partisan debate. Agreement on the necessity of a national defense does not imply consensus on proper funding levels or tactical emphases. Specific weapons systems and even their technical configurations have become political issues, and Congress often extracts a political price for acceding to presidential requests for particular weapons systems. Moreover, it has only approved some systems because enough representatives have been convinced (or have convinced themselves) that the newly approved weapons would be used as bargaining chips in negotiations with the Russians but would never really be deployed.

Moreover, the ability to meet the requisites of national defense is affected by economic and political calculation. British appeasement was dictated in part by the priority given to economic constraints on national security policy.[11] Similarly, the Eisenhower administration's commitment to fiscal stringency dictated a "new look" in defense policy that entailed greater reliance on nuclear weapons.[12] On the other hand, the Kennedy administration's commitment to flexible response and a concomitant defense buildup set the parameters for its domestic economic policy.[13] It soon discovered, though, that an emphasis on flexible response and counterforce targeting made it difficult to hold back subsequent military requests for additional missiles, warheads, and bombers. Defense Secretary McNamara found that discussing the nation's contingency war plans publicly entailed costs. Thus, he shifted back to a doctrine of assured destruction as a way of capping and limiting military requests for more strategic weapons.[14]

In addition, because changes in the level of military expenditures can have implications for other kinds of policies, interests in those other issues can become the basis for positions taken on national security policy. Indeed, political coalitions often form around constellations of policies that address a variety of concerns, including defense. In turn-of-the-century Germany, for example, a naval buildup was one key program of a broad political and social alliance that also supported high tariffs, a nationalistic foreign policy, and an authoritarian domestic government.[15]

Even the basing of weapons has been politically controversial. President Reagan, although a strong proponent of rearmament and military modernization, opposed President Carter's plan to put the MX missile on an underground race track. Two of the president's strongest allies, Republican Senators Paul Laxalt of Nevada and Jake Garn of Utah, had lobbied vigorously against the race track, which would be located in the West, despite their avid support for both the weapon and an increased defense budget. Neither they nor the president wanted to alienate western voters, and so the political impossibility of the race-track option ultimately bounded the Reagan administration's search for a way to deploy the MX missile that it very much wanted.

The scope of politics in the formulation of defense policy is exacerbated when there is no consensus on the necessary means of military strategy and security. When experts disagree, the political considerations always present in the choice of national goals spill over into discussions of means for achieving them. There may be a national consensus on deterrence as an objective of national security policy, but this will not ensure the adoption of any specific policy or policies unless there is a consensus on the means for achieving deterrence. However, such consensus has not existed in the United States.

Rather, two arguments have dominated the American debate on the requisites of deterrence. One finds that deterrence is provided by an assured ability to retaliate with such impact as to rule out the very possibility of attack before it occurs. Thus, nuclear weapons do not represent usable military power; they exist solely to deter. Advocates of assured destruction (or finite deterrence) argue, therefore, that relative numbers of weapons do not matter as long as the nation maintains an adequate ability to retaliate. In contrast, the other side of this debate is articulated by those who believe that the massive retaliation inherent in a doctrine of assured destruction cannot deter because it does not provide a credible response to an array of potential provocations and attacks. Because deterrence requires the ability to wage war and prevail at any level of nuclear escalation, nuclear weapons deter only if they are deployed as usable weapons. The proponents of this logic have thus advocated a counterforce doctrine in order to make the threatened use of nuclear weapons more credible. The implications of these two arguments for national security policy are, of course, very different. Adopting one instead of the other would have clear consequences for determining what kinds and numbers of weapons the United States needs. It is not surprising, therefore, that political considerations can dominate technical ones in the formulation of defense policy.[16]

Yet despite the obvious importance of politics to the formation of national security policy, the debates about the technical and strategic appropriateness of the president's Strategic Defense Initiative are being waged in a political void and do not reflect the array of concerns that underlay the administration's adoption of the program. The president's enunciation of his Strategic Defense Initiative was and is inherently political. It emerged at a politically difficult time for the administration and was intended primarily to solve a domestic political problem.

THE DOMESTIC FIRESTORM OF 1981–82

President Reagan took office in 1981 committed to restoring America's military might. He had campaigned against a Democratic administration that he pilloried for letting America's position in the world decline—for weakness and vacillation and for allowing the United States to be eclipsed by its Soviet enemy. His own defense agenda was wide-ranging, a vast rearmament plan that called for large increases in military spending.[17] Although not everyone in his administration was prepared to embrace the Republican party platform's call for American military superiority, the president argued that rearmament was necessary because of a Soviet "margin of superiority."[18] The president, who had attacked previous arms control agreements, was in no hurry to pursue arms control before his rearmament program was well under way.[19] Indeed, not even interested in a summit meeting with the Soviets, he successfully avoided one during his first term.

Moreover, the new administration's plan included a rhetorical offensive to characterize the Soviet Union as an international brigand. At the president's first news conference, he described the Soviets as being willing to justify as moral anything that furthered their cause. This meant, he continued, that they "reserve unto themselves the right to commit any crime, to lie, [and] to cheat."[20] Three and a half months later, he declared that the West would not contain Communism but transcend it. "We will not bother to denounce it, we'll dismiss it as a sad, bizarre chapter in human history whose last pages are even now being written."[21] Sustained into 1983, this verbal onslaught reached its peak when, in an address to the National Association of Evangelicals, the president characterized the Soviet Union as "the focus of evil in the modern world."[22]

During its first year and a half, the administration coupled its anti-Soviet rhetoric and defense spending increases with public comments about America's own nuclear strategy, an articulation of a United States commitment to a nuclear war-fighting strategy and a recognition of the possibility of nuclear war. In one interview, Sovietologist Richard Pipes, a staff member of the National Security Council, put the chance of nuclear war at 40 percent.[23] Secretary of State Alexander Haig spoke publicly of a NATO plan for low-level nuclear warning explosions.[24] The president himself suggested more than once in late 1981 that tactical nuclear exchanges need not imply full-scale nuclear war.[25] And, in one of the more memorable administration comments, the Undersecretary of Defense for Strategic and Nuclear Forces, T. K. Jones, observed that the United States could fully recover from a nuclear war by emphasizing proper preparation. "If there are enough shovels to go around, everybody's going to make it," he said.[26]

Ironically, the administration was only voicing publicly a long-standing element of American nuclear policy. American nuclear policy had slowly been evolving into a counterforce strategy since the 1950s.

However, these developments were not discussed outside of the defense community, and there existed a separation of targeting strategy and declaratory strategy. As McNamara had learned, one cost of publicly adopting a counterforce doctrine would be the armed services' increasing their lists of targets and demanding more weapons. Whatever the actual war plans, it made good sense to remain publicly committed to a policy of mutual assured destruction and a position that nuclear wars could not be fought. In private, however, it remained necessary to provide for an array of contingencies so that the president had flexibility in case deterrence failed.

The net political effect of the administration's rhetoric and defense policy was to reawaken peace movements on both sides of the Atlantic Ocean. In 1981 and 1982, mass demonstrations overseas protested the planned NATO deployment of American Pershing missiles in Europe.[27] Although European political leaders had previously agreed to accept these weapons, they had also insisted that the United States attempt to negotiate arms reductions with the Soviets. Thus, in 1979, NATO decided to proceed along dual tracks: to negotiate and to begin simultaneously to deploy the Pershings. In 1981, leaders of America's allies pressed the Reagan administration to adopt a negotiating position or proposal that would allow them to withstand domestic political pressures against accepting the Pershings.[28]

However, the administration wanted to avoid arms control negotiations until its rearmament program was in place. It feared that acceptable agreements could not be negotiated without first improving America's defenses, and that negotiations would undercut the momentum for bigger defense budgets. Nonetheless, it responded to pressure from the allies, unveiled a bargaining position, and opened talks in 1981 on intermediate nuclear forces with the Russians. On the other hand, the administration only unveiled its strategic arms control position in May 1982, and discussions with the Russians on these weapons opened the following month.

In addition, the Reagan administration beat a hasty retreat from its war-fighting rhetoric. Nonetheless, it undermined its own effort to sound more pacific. In a conscious attempt to refurbish his image, the president emphasized in the spring of 1982 that "nuclear war cannot be won and must never be fought" and "no one feels more than I the need for peace."[29] Yet the administration defeated its own attempt to alter its public countenance and militaristic diction with its Defense Guidance, a strategic blueprint for budget requests over the next five fiscal years. This document called for developing an American arsenal that could engage in nuclear exchanges with the Soviet Union "over a protracted period." The United States would "prevail" in any nuclear war with the Soviet Union.[30]

Overall, the administration's attempt to change its tone was not enough to quiet the domestic peace movement that its earlier rhetoric had awakened. Although American citizens had been relatively quiescent about strategic issues for years, there burgeoned in the wake of the administration's rhetoric a spontaneous nuclear freeze movement that took Washington by surprise as it gained momentum in 1981 and 1982.[31] "There was Reagan talking about fighting and winning a limited nuclear war and handing out his laundry list of building up every conceivable nuclear weapon because he claimed we were behind the Russians. It brought out the latent anxiety."[32] Even politicians who thought the freeze a bad idea supported at least a partial halt, and their catch phrase, a mutually verifiable nuclear freeze, provided them political cover.[33]

The freeze movement soon dominated the political landscape. But the president opposed a halt to deployments as ingraining Soviet superiority. He lobbied against it in Congress and attacked the movement, which he described as manipulated by people who wanted to weaken America. Arguing that "foreign agents" had instigated the protests, he declared that the freeze advocates were unintentionally "carrying water" for the Soviet Union.[34] Despite the administration's accusations, the movement grew.

In 1982, nuclear freeze initiatives appeared on state and local ballots throughout the country. Administration officials publicly urged voters to reject the freeze.[35] Yet the voters in 8 of 9 states and in 27 of 29 cities and localities approved it. In all, freeze initiatives carried in 41 of 46 contests in 1982. In addition, 205 city councils, 43 county governments, and 11 state legislatures passed freeze resolutions. Movement supporters also argued that the issue had become an important factor in a number of races for office in 1982.[36]

The success of the freeze at the polls in 1982 also affected the 1984 presidential campaign. All but one of the Democratic candidates came out in support of the freeze, and one aspirant actually hoped to parlay his exclusive emphasis on the peace issue into the nomination. Indeed, one way that former Vice-President Mondale finally bested Senator Gary Hart in the primaries was to criticize Hart's delay in backing the freeze and emphasize Hart's flip-flops on the issue.[37]

But in 1982, the administration confronted a problem not just confined to the political left. The domestic debate over nuclear weapons that had begun in 1981 now included the bishops of the Roman Catholic church, who established an Ad Hoc Committee on War and Peace in 1981. In June 1982, the committee released the first public draft of their pastoral letter on war and peace, a document that outlined the bishops' "profound doubts about whether the use or threatened use of nuclear weapons can be truly reconciled with traditional principles of self-defense and just war." The administration responded, publicly and privately, by trying to change the bishops' minds. The White House was worried about the implications of the pastoral letter not only for defense policy but also for broader political reasons. The administration and the bishops had, after all, concurred on such critical issues as tuition tax credits and abortion, and the Republicans needed ethnic blue-collar voters. Now it appeared that the nuclear rearmament plan might make it more difficult to appeal to Catholic voters.

But the administration could not confront the bishops' challenge by vigorously attacking them as it had the freeze movement.[38] Its failure to sway them became clear when, in October 1982, the bishops released a second public draft of their letter in which they advocated a freeze, objected to nuclear war-fighting rhetoric, and hinted they might oppose the MX specifically in a subsequent draft. Once again, the administration attempted to convince the bishops that moral considerations indeed guided American policy, and that it was merely continuing long-standing American strategic nuclear policy.[39] When the bishops released their third draft in May 1983, they did, in fact, soften their stand somewhat. But then they overwhelmingly adopted a final version that restored much of the second draft's language.[40] Over time, the bishops' involvement served to generate a widespread debate on the morality of deterrence and American strategic doctrine. The domestic debate had widened and deepened, for the draft pastoral letters represented a frontal assault on American strategic doctrine by questioning the morality of nuclear weapons and the doctrines associated with their use.

The president thus continued to confront a major political problem. His approach to national security policy was under assault, and his image threatened. When he ran for president in 1980, he had been advised to come across as "peace-oriented." Yet he continued to have a "warmonger" problem.[41] Retreating from his nuclear war-fighting rhetoric had been insufficient; so had been his attempt to go beyond the SALT (Strategic Arms Limitation Talks) agreements he had previously attacked while continuing to distance himself from them. By dubbing his own arms control negotiations START (Strategic Arms Reduction Talks), Reagan tried to emphasize the goal of reducing, rather than just limiting, nuclear weapons. Similarly, the president's arms control proposal, unveiled in May 1982, also reflected the administration's concern with its appearance. By seeking deep cuts in the number of large land-based missiles, the president could maintain that he had outdone the freeze advocates in his attempt to diminish the arsenal of nuclear weapons rather than merely hold it at current levels.[42] But critics quickly dubbed the president's proposals one-sided and, therefore, nonnegotiable, because the administration demanded that the Soviets dismantle most of their land-based missiles without similar concessions.

Thus, the administration's efforts to improve the image of the president and of his party had failed. The Republican party could survive the 1982 recession, the memory of which would fade with economic recovery, but being seen as the party of the bomb could have more lasting consequences. The conjunction of anti-Soviet and nuclear war-fighting rhetoric with large military spending increases and a wary commitment to arms control had generated a major public relations problem.

Moreover, this public relations problem had concrete ramifications for policy making. Despite his general success in vastly increasing the defense budget, Reagan found it very difficult to obtain congressional approval for the MX missile, the cornerstone of his strategic modernization program. The weapon would not only provide the United States with a land-based counterforce capability, but also would strengthen what was seen as an increasingly vulnerable leg of the strategic triad. Yet the president had trouble finding a way to base the MX that neither left it vulnerable nor upset Republican westerners. At the same time, the freeze movement and other opposition to new deployments compounded his difficulties. By the winter of 1982, Congress had rejected two basing schemes proposed by the president and had eliminated funds to produce the MX.[43] Two years into his term, the land-based portion of the president's strategic modernization program appeared to be derailed.

The president attempted to solve this political problem by trying to duplicate his successful strategy for dealing with the Social Security crisis; he named a bipartisan Commission on Strategic Forces in January 1983. He hoped that the Scowcroft Commission, as it became known, would provide a solution to the MX-basing problem that would be politically acceptable to Congress. It became apparent even before the commission submitted its report that it would recommend deploying the MX in existing silos.[44] But because Congress had previously rejected such a plan, it was not clear that it would now accept the commission's recommendation to do what everyone recognized as nothing more than a carefully crafted justification of a political solution to a technical and political problem.

Thus, the president was trying to do two things in his March 23, 1983, address: generate the support he needed for the annual battle over the defense budget and reposition himself on the nuclear issue. His actual proposal, to explore the possibility of a defense against offensive nuclear weapons, caught most nuclear strategists and even many members of the administration by surprise. Although the proposal had not been thoroughly thought through, it had a simple and basic appeal.[45]

The president could now emphasize his passionate commitment to peace and to a world without nuclear weapons. He could out-peacemonger anybody. No one could argue with the simple appeal of relying on defensive weapons rather than offensive ones. Reagan led even his liberal critics to consider the possibility. As columnist Meg Greenfield asked, "Is no such initiative worthy? Is it unfit for contemplation? . . . I wish the status quo nuclear gang would try to improve on Reagan's thought, not merely satirize it. I wish they, too, would think radically."[46] Any who questioned the proposal risked "being labelled tired and negative thinkers."[47] The president's proposal, "an utter masterstroke,"

> has the immediate effect of illuminating the political landscape. We are now beginning to see which of the advocates of American disarmament really want a nuclear-free world and which really want something else—empty moral grandstanding, the promotion of Soviet interests, the abasement of America.
>
> Already it is remarkable how many of these intrepid moralists are not even attracted by the Reagan vision and refuse to entertain it for even a fleeting moment.[48]

Reagan could pursue rearmament and new weapons deployments while he responded to criticism by pointing to his ultimate hope—a world without nuclear weapons. He could turn the Catholic bishops' arguments against his detractors, and he could argue that to reject his vision would be to continue relying on the immoral threat of mass murder.

Yet the president had done little more than elevate and highlight a research effort already underway. The 1980 Republican party platform had called for "vigorous research and development of an effective antiballistic-missile system," and various fringe elements had pressed for space-based defensive systems.[49] A White House position paper of October 1981 proposed expanding defenses in general and in space as well.[50] The Defense Guidance had already called for stepped-up space activities, including those "in support of [the U.S.] right to self-defense," and on September 1, 1982, the Air Force had established a new Space Command.[51] But in March 1983, the president gave these efforts his imprimatur and elevated them to national goals. A politically savvy president took rhetoric already in the air, put it in a form that made it difficult to oppose, and tapped a variety of concerns that resonated with many people. The rest of his administration would now have to defend the president's vision publicly.[52]

THE AFTERMATH

In the ensuing months, the president discovered that he needed to demonstrate some political flexibility to ensure congressional approval for deployment of the MX. The Scowcroft Commission admitted that its recommendation to base the MX in existing silos was a political decision rather than a military one.[53] Yet even this bipartisan recommendation could not ensure congressional support, and the president was pressed to modify his 1982 arms control proposal on strategic weapons in order to secure the MX.[54] And ultimately, the president's congressional victory was assured by the support of representatives who voted for a missile they thought a poor idea. The congressional approval of MX deployment was an act "of well-intentioned people acting out of a combination of their own sense of what it means to be 'responsible'; their own sense of what they needed to do to 'position' themselves politically or lay the groundwork for political advancement; and political fear."[55]

This political trade of the MX for arms control flexibility represented the fashioning of a bipartisan commitment on deployments and arms control.[56] That congressional support was sustained during the 1984 presidential election year despite the collapse of arms control negotiations with the Soviets, who at the end of 1983 walked out of both the talks on intermediate nuclear forces and those on strategic weapons.[57]

Armed with congressional support for his arms control proposals and his strategic modernization program, the president could sit back and wait for the Russians to return to the bargaining table. When they did offer to resume talks, in June 1984, it was SDI, and SDI alone, that they wanted to talk about. The United States insisted that any talks deal with Earth-based weapons as well. The USSR initially refused. But in November 1984, after the president's landslide reelection, the Soviets agreed to return to the bargaining table. By the fall of 1985, it had become clear that they were prepared to negotiate reductions in their land-based forces in exchange for constraints on American development of space-based defensive weapons.

The Soviets' concern with SDI only reinforced the president's position. For it became politically more difficult to attack Star Wars once it became clear that this had brought the Soviets back to the negotiating table. The president, who had not had much with which to negotiate, had moved himself into an enviable position. The challenge of a qualitative arms race concerned the Soviets in a way that a small number of MX missiles did not. Merely articulating his vision had already borne political fruit.

CONCLUSION

The president's enunciation of his vision of a space defense was based on domestic political concerns and had important international political ramifications. He refashioned the domestic political debate over strategic weapons, doctrine, and negotiations. The president did what all successful politicians do—he shifted the focus of debate. He shifted the focus from a troublesome present to a desirable future.[58] In positioning himself as a man with a vision of peace, he lit on a strategy that brought the Soviets back to the bargaining table.

The president's speech generated a national debate; it both scrambled political agendas and recast rhetorical positions. As one senator told his colleagues, "the political landscape has been changed by a speech we all know to be wrong."[59] The administration had complained about the attempt by freeze leaders to mobilize support through simplistic slogans. Since the speech, however, "freezeniks" have accused the supporters of the peace shield of trying to sell the public an appealing unicorn, and they lament that imagery seems more important than reality in a world dominated by television and permanent political campaigning.[60] Moreover, the president's critics have had to find new rationales for their opposition to his newly articulated vision. Leaders of the nuclear freeze movement, for example, have shifted their focus. Explicitly rejecting deterrence altogether, they have made complete nuclear disarmament their goal.[61]

Others, including members of the administration, found that they could support SDI without accepting the president's vision. Few professionals, including those who ran the Pentagon Strategic Defense Initiative Office, thought that "leakproof protection" was "possible or even necessary."[62] As even Lieutenant General James Abrahamson, the Pentagon manager of SDI, presciently admitted, "Nothing is perfect, not even the [space] shuttle. A perfect astrodome defense is not a realistic thing."[63] Many initial believers eventually backpedaled.[64] Those who did not buy the whole package could still argue, however, that even if Reagan's ultimate goal proved "unattainable," a "mirage," some degree of defense would "complicate the attacker's calculations" and thus improve deterrence.[65] In other words, this alternative formulation offered grounds on which the president's proposal could be supported and marketed. The catchwords, inside and outside the administration, held that SDI would "enhance deterrence" and "strengthen stability."[66]

The difficulty of opposing the president has brought an array of unlikely supporters flocking to his SDI bandwagon. Although few strategists accept the president's vision of its potential, most have found other reasons to support it.[67] Minimally, even its critics recognize the negotiating possibility that SDI has offered; they simply urge the president to bargain it away in an agreement with the Russians. Even President Carter's Secretary of State, Cyrus Vance, who opposes SDI deployment, has endorsed research on space weapons as "prudent" and admits that "no president for the last 12 years has been in a stronger position to make progress on [arms control] than is President Reagan."[68]

At the same time, the inability to attack SDI directly has led its critics to find new ways to frame their opposition. For example, they emphasize the importance of maintaining the ABM (anti-ballistic missile) treaty and not engaging in any program that would violate it. On the other hand, some of SDI's supporters would be happy to violate it in order to get rid of the treaty. Thus, the Star Wars debate may test the degree to which the nation has accepted the ABM treaty, just as battles over the administration's domestic policies have tested how ingrained various Great Society and New Deal programs have become.

Debates about military strategy and national security policy can be carried on without reference to politics, but their formulation and adoption necessarily involve politics.[69] And the politics is not just international, but also domestic, including electoral and party politics. The adoption of a direction for national security policy thus involves more than a calculation of optimal defense policy. Positioning for elections and domestic bargaining, as well as for international negotiations, is also important. And public statements, like bids at an auction, are first offers—not rock-bottom positions.

Thus, proclamations about nuclear strategy and the direction of security policy are about more than just the technical means to the sole end of maintaining the territorial integrity of the state. Yet even if no other objectives for national security policy existed, the nature of an optimal nuclear strategy and doctrine would remain contested. For unlike economists discussing trade barriers, nuclear strategists have reached no consensus about the requisites of deterrence.

Rather, theorists remain divided about the requirements of deterrence in an era of mutual vulnerability. For some, deterrence requires survivable forces that can inflict unacceptable retaliatory damage. For others, such a doctrine is not credible, and deterrence requires an ability actually to wage nuclear war. Yet this debate has been largely irrelevant to the recent history of doctrine and deployments, because American policy has included counterforce options and capabilities for more than two decades. Nonetheless, there is no unanimity about the means necessary to achieve deterrence. Indeed, there is no way empirically to resolve this dispute.[70]

Even consensus on both the desirability of deterrence and the means of achieving it might not end the national debate on strategic policy. Advocates of assured destruction, for example, define stabilizing actions as those that reduce the value of a first strike while increasing the value of a second strike. They agree not only on the desirability of stability but even about which weapons systems are stabilizing and which destabilizing. Yet they can still advocate different routes to stability. For anything that increases one nation's value of a second strike decreases the other's value of a first strike. That is, stability can be improved either by degrading the other side's value of a first strike or accepting the degradation of one's own potential for striking first. Some globalists are prepared to eschew a nation-centered view and accept the degradation of their own nation's forces for the sake of planetary welfare. On the other hand, it is not surprising that nationalists prefer to achieve stability by reducing the other nation's first-strike potential rather than their own, and that nations react to the reduction of their own forces' efficacy by the other with destabilizing actions. In short, the acceptance of stability does not ensure agreement on acceptable and unacceptable deployments.

Moreover, as the history of economic policy makes clear, scholarly consensus on means–ends relationships can still be ignored by politicians when they cannot square politics and economics, but where the political imperative is strong. Congress ignored the opposition of more than one thousand economists when it imposed the Smoot–Hawley tariff in the midst of the depression. Similarly, the recent Senate adoption of the Gramm–Rudman plan demonstrates the lengths to which politics can drive policy. Confronted by the need to extend the government debt ceiling above the symbolically important figure of two trillion dollars, the Senate adopted along with it a plan to reduce the deficit by 1991. Few who voted for the proposal, improvised at the last minute without any hearings, thought it a good idea. Few knew how it would work, or if it was viable or even constitutional. But driven by political maneuvering for the 1986 election and the Senators' fear of the political costs of not voting to reduce the deficit, the proposal carried. As an aide to one Senate Democrat said:

> If you had a dime for every time this proposal has been called a turkey in both parties' caucuses, you'd be a rich person. It's the politics of the issue that are driving it, not the merits, and therefore people here don't care what the economists think. The Democrats are having a 'cover your backside' reaction—they feel the country would cheer a balanced budget. . . . What this is all about is a fight over control of the Senate.[71]

Electoral concerns, political worries, and the need to cover one's backside are evident in defense policy as well. The president and his advisers, aware of this, have exploited all three successfully since the latter part of his first term. In the belief that arms-control policy is unique in the quality of its discourse, one administration official recently lamented that "the real casualty in this area comes from discussions of things that have no bearing on reality." Citing a leakproof defense shield as one example of the intrusion of the chimerical into disputes over arms-control policy, he argued that "this wouldn't happen in any other area of public policy. It may be good for the polls, and for throwing people for a loop in a debate, but ultimately it's a disservice. It's very costly in terms of serious discussion of the subject."[72]

In 1947, Senator Arthur Vandenburg, Chairman of the Senate Foreign Relations Committee, told President Truman and his advisors that to obtain support for their aid plans for Greece and Turkey, they would have to scare the hell out of the American people. By the late 1970s, this dictum no longer held true. New nuclear deployments now had to be coupled with arms control negotiations. NATO had enshrined this shift in its dual-track decision of 1979. The Reagan administration was slow to recognize the implications of this change for European and American politics. New American weapons deployments had to be coupled with the perception of an American commitment to arms control and linked to a vision of peace, not one of waging nuclear war.

Nuclear strategists are engaged in a political debate, and their arguments, like those of economists, are employed or ignored as suits political needs. National security policy in open societies is not a depoliticized issue being handled by experts without political considerations. The nature of deterrence remains essentially contested. In such a world, politics will determine policy as much as science.

Scholars engaged in a search for viable strategic doctrines must be aware of the political as well as military requisites of national security policy. Policy makers confront domestic and international political demands and must be sensitive to them. Scholars who address such policy makers must be aware of these considerations. As the war fighters discovered, scaring your own people and those of your allies is not a way to enshrine your strategic vision. Increases in defense spending, and especially the deployment of offensive nuclear systems, must be linked to some vision other than waging nuclear war. As the finite deterrence theorists discovered, the threats of mass murder and nuclear apocalypse are insufficient visions for sustaining national security policy.

NOTES

1. The characterization of the president is from an editorial, "Supply Side and Star Wars," *Baltimore Sun*, April 3, 1983, p. K4.

2. "Address to the Nation on Defense and National Security, March 23, 1983," in *Public Papers of the Presidents of the United States: Ronald Reagan, 1983*, book 1, pp. 437–43.

3. "Supply Side and Star Wars," *Baltimore Sun*, April 3, 1983, p. K4.

4. Public opinion is also responding to SDI as a political rather than technical issue. Questions about SDI generate partisan responses, with Republicans supporting it and Democrats opposing it. In addition, the symbolic underpinnings of the debate can be seen in the ways that the polling results shift with changes in the wording of questions. See Barry Sussman, "On 'Star Wars,' It All Depends on How You Ask the Question," *Washington Post National Weekly Edition*, November 25, 1985, p. 37.

5. Politics is how a society goes about determining societal objectives. Indeed, political systems are distinguishable by the ways in which they do or do not aggregate individual preferences into social choices and by the respective weights they attach to the preferences of different individuals.

6. For a discussion of politics and science, see Max Weber, "Politics as a Vocation," and "Science as a Vocation," in *From Max Weber: Essays in Sociology*, ed. H. H. Gerth and C. Wright Mills (New York: Oxford University Press, 1946), pp. 77–128, 129–56.

7. Allan M. Brandt, *No Magic Bullet: A Social History of Venereal Disease in the United States Since 1880* (New York: Oxford University Press, 1985).

8. Indeed, SDI has not attracted wide support in the Pentagon, not even the Air Force. As one official put it, "The Air Force is concerned their budget is going to get chewed up." Money for lasers is money that could go to other projects. See Fred Kaplan, "Reagan's 'Star Wars' at key stage," *Boston Globe*, May 13, 1984, p. 1.

9. Politicians worry, for example, about the impact of economic policies on their bases of support. Some, therefore, fear unemployment; others dread inflation. Thus, presidents and congressional representatives may favor and adopt policies that serve their own interests but that are not optimal from the perspective of the performance of the economy as a whole. This is the basis for arguments about the existence of a political business cycle.

10. For one compilation of incoherence in Reagan strategic policy, see Robert Jervis, *The Illogic of American Nuclear Strategy* (Ithaca, N.Y.: Cornell University Press, 1984). The same kind of argument can be made about the peculiar combination of economic doctrines embedded in Reaganomics, which in 1981 welded monetarism and supply-side tax cuts.

11. Paul Kennedy, "The Tradition of Appeasement in British Foreign Policy, 1865–1939," *British Journal of International Studies* 2 (October 1976), pp. 195–215; and Paul Kennedy, "Strategy versus Finance in Twentieth-century Britain," *International History Review* 3 (1981), pp. 45–61.

12. Glenn H. Snyder, "The 'New Look' of 1953," in *Strategy, Poli-*

tics, and Defense Budgets, eds. Warner R. Schilling, Paul Y. Hammond, and Glenn H. Snyder (New York: Columbia University Press, 1962), pp. 379–524.

13. Amy E. Davis, "The Dangers of an Upturn" (Department of History, Columbia University, November 1985).

14. For the evolution of strategic thought and policy, see Lawrence Freedman, *The Evolution of Nuclear Strategy* (New York: St. Martin's Press, 1981); Fred Kaplan, *The Wizards of Armageddon* (New York: Simon and Schuster, 1983); and John Lewis Gaddis, *Strategies of Containment: A Critical Appraisal of Postwar American National Security Policy* (New York: Oxford University Press, 1982).

15. See, for example, Eckart Kehr, *Economic Interest, Militarism, and Foreign Policy: Essays on German History* (Berkeley: University of California Press, 1977).

16. For a discussion contrasting assured destruction and flexible response, see Robert Jervis, "Why Nuclear Superiority Doesn't Matter," *Political Science Quarterly* 94 (Winter 1979–80), pp. 617–33. Still more trenchant criticisms of flexible response can be found in Jervis, *The Illogic of American Nuclear Strategy*, and Robert Jervis, "Strategic Theory: What's New and What's True," in this volume. One example of a war-fighting doctrine can be found in Colin Gray, "Nuclear Strategy: The Case for a Theory of Victory," *International Security* 4 (Summer 1979), pp. 54–87.

17. The administration generated estimates suggesting that the Soviets had outspent the United States by almost half a trillion dollars over a decade. See Michael R. Gordon, "Rubles for Defense—Are the Soviets Really Outspending the Pentagon?" *National Journal* 13, no. 15 (April 11, 1981), pp. 601–604.

18. "The President's News Conference, March 31, 1982," in *Public Papers of the Presidents of the United States: Ronald Reagan, 1982*, book 1, p. 399.

19. Richard Halloran, "Weinberger Calls Nuclear Arms Buildup Prerequisite to Talks," *New York Times*, November 4, 1981, p. A6. For a description of the administration's tortuous discussions on whether to adhere to SALT II, an accord that the president had labeled "fatally flawed," and the discussions underlying the administration's arms control proposals, see Strobe Talbott, *Deadly Gambits: The Reagan Administration and the Stalemate in Nuclear Arms Control* (New York: Alfred A. Knopf, 1984).

20. "The President's News Conference, January 29, 1981," in *Public Papers of the Presidents of the United States: Ronald Reagan, 1981*, p. 57. Later in the year, the president replied to a question about a softening in his attitude by reiterating this thought and concluded, "Now, I didn't set out to talk harshly about them. I just told the truth." See "The President's News Conference, December 17, 1981," in *Public Papers of the Presidents of the United States: Ronald Reagan, 1981*, p. 1167. Almost two years later, Reagan was still trying to extricate himself from these comments. See "Reagan on Lenin," *New York Times*, January 22, 1983, p. 7.

21. "Excerpts from the President's Address," *New York Times*, May 18, 1981, p. B7.

22. "Remarks at the Annual Convention of the National Association of Evangelicals in Orlando, Florida, March 8, 1983," in *Public Papers of the Presidents of the United States: Ronald Reagan, 1983*, book 1, p. 363.

23. Charles Fenyvesi, "The Man Not Worried by the Bomb," *Washington Post*, April 11, 1982, p. B1.

24. Bernard Gwertzman, "Allied Contingency Plan Envisions a Warning Atom Blast, Haig Says," *New York Times*, November 5, 1981, p. 1.

25. The president said, "I could see where you could have the exchange of tactical weapons against troops in the field without it bringing either one of the major powers to pushing the button." "Remarks and a Question-and-Answer Session at a Working Luncheon with Out-of-Town Editors, October 16, 1981," in *Public Papers of the Presidents of the United States: Ronald Reagan, 1981*, pp. 956–57. The President's clarifications did not help; see "The President's News Conference, November 10, 1981," in *Public Papers of the Presidents of the United States: Ronald Reagan, 1981*, p. 1033.

26. Robert Scheer, *With Enough Shovels: Reagan, Bush and Nuclear War* (New York: Random House, 1982), p. 18. That nuclear war could not be survived was one of the ideas that supporters of war-fighting scenarios sought to dispel. For one example, see Edward Teller, "Dangerous Myths about Nuclear Arms," *Reader's Digest*, November 1982, pp. 139–44.

27. The European protest movement was linked in part to Reagan rhetoric. See Senate Committee on Foreign Relations, *NATO Today: The Alliance in Evolution*, 97th Cong., 2d sess., April 1982, Doc. no. 97–31, p. 79.

28. See Henry Tanner, "Haig, in Europe, Is Pressed to Agree to Nuclear Arms Talks," *New York Times*, May 3, 1981, p. 18; and John Vinocur, "Reagan Writes Letter to Brezhnev about Talks on Missiles for Europe," *New York Times*, May 5, 1981, p. 1. Also see Talbott, *Deadly Gambits*, ch. 2.

29. "Subtle Shifts in Image for the President," *New York Times*, April 20, 1982, p. A22. The quotations can be found in Leslie H. Gelb, "Reagan Says That Peace Depends on Support for U.S. Arms Buildup," *New York Times*, April 18, 1982, p. 1.

30. Richard Halloran, "Pentagon Draws Up First Strategy for Fighting a Long Nuclear War," *New York Times*, May 30, 1982, p. 1. Also see, Richard Halloran, "Weinberger Denies U.S. Plans for 'Protracted' War," *New York Times*, June 21, 1982, p. 5; "U.S. Aide Discloses New Policy on Russians," *New York Times*, June 17, 1982, p. A23; Richard Halloran, "Weinberger Defends His Plan On a Protracted Nuclear War," *New York Times*, August 10, 1982, p. 8; Richard Halloran, "Weinberger Angered by Reports on War Strategy," *New York Times*, August 24, 1982, p. B8; and Richard Halloran, "After Detente, the Goal Is to Prevail," *New York Times*, September 23, 1982, p. B16. Also see Richard Halloran, "Weinberger Said to Offer Reagan Plan to Regain Atomic Superiority," *New York Times*, August 14, 1981, p. 1.

31. Paul Boyer, "From Activism to Apathy: The American People and Nuclear Weapons, 1963–1980," *Journal of American History* 70 (March 1984), pp. 821–44.

32. Professor Henry Kendall, quoted in Fox Butterfield, "Anatomy of the nuclear protest," *New York Times Magazine*, July 11, 1982, p. 17.

33. Mark M. Lowenthal and Judith A. Freedman, "Congress and

the Nuclear Freeze," in *Congress and Foreign Policy, 1982,* House Committee on Foreign Affairs (1983).

34. "Address Before a Joint Session of the Tennessee Legislature in Nashville, March 15, 1982," in *Public Papers of the Presidents of the United States: Ronald Reagan, 1982,* book 1, p. 299; Bernard Gwertzman, "President Rejects A Nuclear Freeze But Calls For Cuts," *New York Times,* April 1, 1982, p. 1; Hedrick Smith, "The Nuclear Freeze," *New York Times,* April 1, 1982, p. A23; Steven R. Weisman, "Reagan Calls on Catholics in U.S. To Reject Nuclear Freeze Proposal," *New York Times,* August 4, 1982, p. 1; Judith Miller, "President Lobbies Against Freeze On Eve of Vote on Nuclear Arms," *New York Times,* August 5, 1982, p. 6; "President Says Foes of U.S. Have Duped Arms Freeze Group," *New York Times,* October 5, 1982, p. A22; "The President's News Conference, November 11, 1982," in *Public Papers of the Presidents of the United States: Ronald Reagan, 1982,* book 2, p. 1455; Leslie Maitland, "Sources Are Cited for Charge of Soviet Tie to Arms Freeze," *New York Times,* November 13, 1982, p. 7; Judith Miller, "President Says Freeze Proponents May Unwittingly Aid the Russians," *New York Times,* December 11, 1982, p. 9. But see also, Judith Miller, "Soviet Role in Freeze Movement Found Minor," *New York Times,* December 10, 1982, p. A15.

35. In addition to the president's numerous attacks mentioned above, see Richard Halloran, "Weinberger Bids Voters Reject a Nuclear Freeze," *New York Times,* October 29, 1982, p. 22; and Judith Miller, "The Brothers Lehman: Battling Nuclear Freeze," *New York Times,* November 8, 1982, p. B8.

36. Material in this paragraph is drawn from William R. Anderson, "Freeze proponents send Washington a message, and Washington replies," *National Journal* 14, no. 45 (November 6, 1982), pp. 1875, 1910. See also Neal R. Peirce and William R. Anderson, "Nuclear Freeze Proponents Mobilize on Local Referenda, House Elections," *National Journal* 14, no. 38 (September 18, 1982), pp. 1602–1605.

37. Robert Pear, "7 Presidential Contenders Debate Arms Control," *New York Times,* October 14, 1983, p. B5; Robert Pear, "Democratic Candidates Disagree over Adequacy of Nuclear Force," *New York Times,* November 6, 1983, p. 26; and Elizabeth Drew, "A Political Journal," *The New Yorker,* April 2, 1984, pp. 65–66. For the 1984 election campaign, see William A. Henry, *Visions of America: How We Saw the 1984 Election* (Boston: Atlantic Monthly Press, 1985); Peter Goldman and Tony Fuller, *The Quest for the Presidency 1984* (New York: Bantam Books, 1985); and Elizabeth Drew, *Campaign Journal: The Political Events of 1983–1984* (New York: Macmillan Publishing Company, 1985).

38. Declining to criticize the bishops' letter when it was finally approved, the president argued only that the administration, like the bishops, wanted world peace. See "Question-and-Answer Session with Reporters on Domestic and Foreign Policy Issues, May 4, 1983," in *Public Papers of the Presidents of the United States: Ronald Reagan, 1983,* book 1, pp. 635–36.

39. Richard Halloran, "Bishops Joining Nuclear Arms Debate," *New York Times,* October 4, 1982, p. B6; Marjorie Hyer, "Bishops Strengthen Arms Draft," *Washington Post,* October 26, 1982, p. 1; David Wood, "Proposed Letter by Bishops Asks Nuclear Arms Freeze," *Los Angeles Times,* October 26, 1982,

p. 1; Richard Halloran, "U.S. Tells Bishops Morality Is Guide on Nuclear Policy," *New York Times*, November 17, 1982, p. 1; Richard Halloran, "Minuet with Catholic Bishops over Nuclear War," *New York Times*, December 16, 1982, p. B16. For an emphatic assertion of continuity, see U.S. Secretary of Defense Caspar W. Weinberger, *Annual Report to the Congress*, February 1, 1983, p. 51.

40. The National Conference of Catholic Bishops of the U.S. adopted a Pastoral Letter on War and Peace in May 1983 by a vote of 328 to 9. For an excellent discussion, see Bruce M. Russett, "Ethical Dilemmas of Nuclear Deterrence," *International Security* 8 (Spring 1984), pp. 36–54. Also see Kenneth A. Briggs, "Roman Catholic Bishops Toughen Stance against Nuclear Weapons," *New York Times*, May 3, 1983, p. 1; Kenneth A. Briggs, "Bishops Endorse Stand Opposed to Nuclear War," *New York Times*, May 4, 1983, p. 1; Paula Herbut, "U.S. Catholic Bishops Adopt Anti-Nuclear Pastoral Letter," *Washington Post*, May 4, 1983, p. 1.

41. Elizabeth Drew, "A Political Journal," *The New Yorker*, May 9, 1983, pp. 48, 70. Also see Elizabeth Drew, *Portrait of an Election: The 1980 Presidential Campaign* (New York: Simon and Schuster, 1981).

42. In one address, the president emphasized that the arms control proposals put forward by his administration went "far beyond the sterile idea of a freeze." See "Remarks at the Centennial Meeting of the Supreme Council of the Knights of Columbus in Hartford, Connecticut, August 3, 1982," in *Public Papers of the Presidents of the United States: Ronald Reagan, 1982*, book 2, p. 1013.

43. For an analysis of the administration's failure at the end of 1982, see Steven R. Weissman, "White House Setback on MX Had Diverse Roots," *New York Times*, December 17, 1982, p. 21. Also see Richard Halloran, "The Long Search For MX Base Plan," *New York Times*, May 12, 1983, p. A19; and Robert C. Gray, "Congress, Arms Control, and Weapons Modernization," in *Congress and Foreign Policy, 1983*, House Committee on Foreign Affairs (1984), pp. 86–109.

44. Hedrick Smith, "Basing of MX in Minuteman Silos Expected as Panel's Interim Plan," *New York Times*, March 23, 1983, p. 1.

45. Leslie H. Gelb, "Aides Urged Reagan to Postpone Antimissile Ideas for More Study," *New York Times*, March 25, 1983, p. 1. When asked how carefully the president's proposal had been thought through, one administration official said, "I'd put it somewhere between a cocktail napkin and a menu." See Elizabeth Drew, "A Political Journal," *The New Yorker*, May 9, 1983, p. 49.

46. Meg Greenfield, "Calling 'Buck Rogers,' " *Newsweek*, April 4, 1983, p. 88.

47. Elizabeth Drew, "A Political Journal," *The New Yorker*, May 9, 1983, p. 50.

48. Joseph Sobran, "Who Can Object to Obsolescing Nukes?" *Washington Times* April 4, 1983, p. C–2.

49. William J. Broad, "Reagan's 'Star Wars' Bid: Many Ideas Converging," *New York Times*, March 4, 1985, p. 1. Also see Patrick E. Tyler, "How Edward Teller Learned to Love the Nuclear-pumped X-ray Laser,"

Washington Post, April 3, 1983, p. D1. In the wake of Reagan's speech, White House aides argued that the president had thought of space-based defenses much earlier, perhaps even when he was governor of California. See David Hoffman and Lou Cannon, "President Overruled Advisers on Announcing Defense Plan," *Washington Post*, March 26, 1983, p. 1, and "Rethinking the Unthinkable," *Newsweek*, April 4, 1983, pp. 16–18.

50. "Reagan for the Defense," *Time*, April 4, 1983, pp. 8–14, 19.

51. Richard Halloran, "U.S. Plans Big Spending Increase for Military Operations in Space," *New York Times*, October 17, 1982, p. 1. See the table on strategic defense research budgets in Pat Towell, "Reagan, Critics Square Off on Space Weaponry," *Congressional Quarterly*, April 14, 1984, p. 838. Also see Robert Jastrow, "The New Soviet Arms Buildup in Space," *New York Times Magazine*, October 3, 1982, p. 100; Richard Halloran, "U.S. to Increase Military Funds For Space Uses," *New York Times*, September 29, 1982, p. 1; Philip M. Boffey, "Pressures Are Increasing for Arms Race in Space," *New York Times*, October 18, 1982, p. 1; John Noble Wilford, "Military's Future in Space: A Matter of War or Peace," *New York Times*, October 19, 1982, p. 1.

52. As one skeptical official put it, "This is the President's program. We can't tell the president that he's got a nutty idea." See Fred Kaplan, "A War in the Stars?" *Boston Globe*, October 16, 1983, p. 25.

53. Steven V. Roberts, "Leader of MX Panel Says Politics Played Key Role in Decision," *New York Times*, April 19, 1983, p. 1.

54. Steven R. Roberts, " 'Window of Opportunity' in Debate on MX Missile," *New York Times*, May 15, 1983, p. 4E; Francis X. Clines, "Is It Bipartisanship or Is It Symbiosis?" *New York Times*, May 16, 1983, p. A14; Leslie H. Gelb, "Target of 'Flexibility' Is Congress, Not the Kremlin," *New York Times*, June 12, 1983, sec. 4, p. 1.

55. Elizabeth Drew, "A Political Journal," *The New Yorker*, June 20, 1983, p. 39.

56. When William Clark left the position of national security adviser, he warned his colleagues not to allow the upcoming presidential campaign to jeopardize bipartisan support for the president's arms control initiative; see "Unusual and Unlikely," *New York Times*, October 14, 1983, p. A18.

57. The congressional rationale for proceeding with the MX in the spring of 1984 was the opposite of its reasoning in the spring of 1983. See Elizabeth Drew, "A Political Journal," *The New Yorker*, May 28, 1984, p. 119. Also see Robert C. Gray, "Congress, MX, and Arms Control," in *Congress and Foreign Policy, 1984*, House Committee on Foreign Affairs (1985), pp. 48–69.

58. When things are going well, as with the economy in 1984, politicians emphasize current success and avoid discussing the future. Such political strategies parallel a lawyer's rule, "When the law is against you, argue the facts. When the facts are against you, argue the law." See Paul Dickson, *The Official Rules* (New York: Dell Publishing, 1978), pp. 103–104.

59. *Congressional Record*, 99th Cong., 1st sess., vol. 131, no. 72, June 4, 1985, p. S7353.

60. "Talk of the Town," *The New Yorker*, November 18, 1985, pp. 39–40.

61. Mark Hertsgaard, "What Became of the Freeze?" *Mother Jones*, June 1985, pp. 44–7. Freeze leaders, not unlike others who no longer find themselves at the heart of national discourse, blame the media for their disappearance from the political scene.

62. Brad Knickerbocker, "The debate over strategic defense," *Christian Science Monitor*, September 17, 1984, p. 1; Brad Knickerbocker, "Writing a new and contentious script for nuclear deterrence," *Christian Science Monitor*, September 21, 1984, p. 1; R. Jeffrey Smith, "Weapons Bureaucracy Spurns Star Wars Goal," *Science* 224 (April 6, 1984): 32–4. See also Admiral James D. Watkins, "The Strategic Defense Initiative: Keeping Teeth in the Policy of Deterrence," *Defense 85*, March 1985, p. 14.

63. R. Jeffrey Smith, "Star Wars Chief Takes Aim at Critics," *Science* 225 (August 10, 1984): 600.

64. Charles Krauthammer, "Will Star Wars Kill Arms Control?" *The New Republic*, January 21, 1985, p. 12.

65. Henry A. Kissinger, "Should We Try to Defend against Russia's Missiles?" *Washington Post*, September 23, 1984, p. C–8. See also Zbigniew Brzezinski, "From Arms Control to Controlled Security," *Wall Street Journal*, July 10, 1984, p. 36; Bill King, "Brzezinski supports Reagan on 'star wars' " *Washington Times*, January 21, 1985, p. 4; Zbigniew Brzezinski, Robert Jastrow, and Max M. Kampelman, "Defense in Space Is Not 'Star Wars,' " *New York Times Magazine*, January 27, 1985, p. 28.

66. See, for example, U.S. President, *The President's Strategic Defense Initiative*, January 1985, pp. i, 3, 7, 9; see also U.S. Department of State, Bureau of Public Affairs, *The President's Strategic Defense Initiative*, March 1985. Critics complained that this alternative formulation was "not what the President has been talking about." See William Proxmire, "Why Jastrow, Brzezinski and Kampelman are Wrong on 'Star Wars,' " *Congressional Record*, January 29, 1985, p. S737.

67. On the Senate floor, Senator J. Bennett Johnston, Jr. (Democrat, Louisiana) asked if "anybody of any substance believe[s] that speech, believe[s] the essential gravemen [sic] of that speech? The answer . . . is absolutely not. I have not heard a serious scientist yet who has said you could put an astrodome over the United States. . . . Is there anybody here . . . who will get up on this floor and say the President was right that we have any possibility of making nuclear weapons obsolete? Of course not. Of course not." *Congressional Record*, 99th Cong., 1st sess., vol. 131, no. 72, June 4, 1985, p. S7353. For the array of reasons provided in support of SDI, see the list provided in Charles L. Glaser, "Do We Want the Missile Defenses We Can Build?" *International Security* 10 (Summer 1985), pp. 25–57. Some have even reraised the old arguments of supporting SDI to deal with small nuclear powers. See Michael D. Intriligator, "What 'Star Wars' Is Intended to Prevent," letter to the editor, *New York Times*, March 14, 1985, p. A26. The need to deal with small nuclear powers had been invoked as the cover story for McNamara's decision in the middle 1960s to deploy an ABM. See Morton H. Halperin, "The Decision to Deploy the ABM: Bureaucratic and Domestic Politics in the Johnson Administration," *World Politics* 25 (October 1972), pp. 62–95.

68. Bill Outlaw, "Vance backs 'star wars,' flays deployment," *Washington Times*, February 5, 1985, p. 6B.

69. In analyzing the sources of military doctrine, a consideration of political factors is not all that should be added to the more typically cited elements of technology and balance of power. See Barry R. Posen, *The Sources of Military Doctrine: France, Britain, and Germany Between the World Wars* (Ithaca, N.Y.: Cornell University Press, 1984); and Jack L. Snyder, *The Ideology of the Offensive: Military Decision Making and the Disasters of 1914* (Ithaca, N.Y.: Cornell University Press, 1984).

70. In other policy domains, it is possible to experiment in order to assess different arguments. This luxury does not exist in the nuclear world. Robert Jervis argues that the inability to resolve issues empirically enlarges the role of beliefs. See *The Illogic of American Nuclear Strategy*, pp. 37–40. In "Strategic Theory: What's New and What's True," however, Jervis does distinguish valid from invalid ideas and bemoans the incentives for academics to say something original even if false. He also discusses why we should expect intellectual progress to be slight and slow. Yet the academic pressures he mentions do not create dissensus in other fields, and debates about the requisites of deterrence will continue because others do not agree with his specific characterizations of valid and invalid ideas.

71. Elizabeth Drew, "A Reporter in Washington," *The New Yorker*, November 5, 1985, p. 142.

72. Elizabeth Drew, "Letter from Washington," *The New Yorker*, October 27, 1986, p. 134. In addition to a leakproof defense shield, the official's list of unreal elements in discussions of arms control also included the president's offer to share SDI technology with the Soviets and the discussion at the Reykjavik summit about the elimination of all ballistic missiles.

CHAPTER 7

Has Soviet Nuclear Strategy Changed?

BENJAMIN S. LAMBETH

INTRODUCTION

Nearly two decades ago, Bernard Brodie remarked that an ingredient often desperately lacking in American strategic policy was "a thorough knowledge of the special idiosyncracies" of the "adversary . . . as a complex of people organized as a state with distinctive political institutions."[1] That sentiment was plainly out of vogue at a time when cost-effectiveness and narrow quantitative analysis remained at the height of their dominance over United States defense policy. Yet it was far from being an isolated lament. On the contrary, there have been repeated complaints voiced over the years that American defense planning has been overly abstracted from political reality, consistent within its own largely technical terms of reference, but insufficiently responsive to those distinctive stylistic traits of the enemy against whom our weapons and strategies have primarily been directed.

As early as 1958, well before the Soviet military posture had reached anything approaching its current threat dimensions, Raymond Garthoff pointed out that any effort to understand "the strategic thought and doctrine of an alien military culture" required that one first "escape the confines of one's own implicit and unconscious strategic concept."[2] Two decades later, Colin Gray rued our persistent failure to heed such advice when he noted that "most American strategic thinkers have always *known* that there was a uniquely 'Soviet way' in military affairs, but somehow that realization was never translated from insight into constituting a serious and enduring factor

This paper was prepared for presentation at a conference on "The Calculus of Terror: Nuclear Strategy and Its Discontents," sponsored by the UCLA Project on Politics and War, Villa Serbelloni, Bellagio, Italy, December 9–13, 1985.

influencing analysis, policy recommendation, and war planning."[3] More often than not, our military planning has instead been based on what Stanley Hoffmann once called an image of the world as a mere projection of American rationality, in which "opponents are supposed either to reason like Americans or to be in need of education bringing them to this level."[4]

It is a Hobson's choice, however, whether Western defense interests are less well served by assumptions that disregard the doctrine and strategy of the enemy altogether or by those that, through excessive optimism or wishful thinking, misconstrue the motivations underlying his behavior. In the past several years, a significant body of opinion has emerged in the West that holds that the Soviet Union, after two decades of adherence to a belief that it makes good sense to plan for victory in case of nuclear war, has lately taken a more benign turn in its military policy. In its most immodest form, this viewpoint was recently conveyed in a sweeping pronouncement by former United States Defense Secretary Robert McNamara, who accepted without challenge a number of statements by Brezhnev and other Soviet officials and used these as a basis for asserting that those tough sounding Soviet writings of the 1960s, which "were used so devastatingly by opponents of nuclear arms control," were now "badly out of date" and had been rendered irrelevant by what he proclaimed to be a major "doctrinal shift" reflected in Brezhnev's soothing disavowal of any Soviet intent to acquire a nuclear war-fighting capability.[5]

A similar view has also been propounded (albeit with more seemly circumspection) by some well-established professional students of Soviet military affairs. David Holloway, for example, has written that "there are signs that Soviet thinking about nuclear war has changed since the mid-1970s." In his assessment, Brezhnev's statements since the late 1970s have introduced "a new element into Soviet military doctrine" and have rendered "aspects of Sokolovskii's *Military Strategy* outdated."[6] Relatedly, Dimitri Simes has remarked that "a modern nuclear war-winning orientation has . . . become less dominant in Soviet national security philosophy."[7] This image of a Soviet shift toward nuclear moderation has been given the most widely noted expression, however, by Raymond Garthoff, who argues that increasingly since the mid-1970s, the Soviets have come to recognize "that the need to avoid a nuclear war can best be served by prudent actions within a framework of mutual strategic deterrence" based on "assured mutual retaliatory capability."[8]

In my own view, Garthoff is absolutely correct to insist that Soviet leaders are as awed by the specter of nuclear war as anybody else. He is on equally solid ground in pointing out that whatever image of toughness Soviet military writings may continue to project, the Soviet leadership—both uniformed and civilian—harbors nothing but respect for what such a war would mean for the continued livelihood of the Soviet state. He has performed a valuable service by pointing out that although Soviet military doctrine may offer important guidelines for Soviet force development, it scarcely reflects any underlying confidence on the part of the leadership about the prospects of the Soviet Union's actually winning a nuclear war—at least in any politically meaningful sense.

Yet he is quite off the mark, it seems to me, in his effort to prove that Soviet leaders have at last accommodated to the Western concept of mutual deterrence as a desirable *regulator* of the superpower balance. True enough, they accept the *existence* of mutual deterrence, simply because it is an undeniable fact of life. But there is nothing I know of—including the material Garthoff cites—that would indicate any Soviet conviction that the mutual part of the situation is a preferred state of affairs to be maintained into perpetuity. For them, it is the United States, not the Soviet Union, that needs to be deterred. The Soviet Union is, by definition, a "peace-loving" state. When Soviet spokesmen assert that the military balance makes it impossible for *any* state to escape a retaliatory rebuff in response to an attack, it takes little reading between the lines to see that it is the United States, not the Soviet Union, that they have in mind. They most certainly are not voicing contentment that *Soviet* security is vouchsafed because the *United States* can successfully retaliate against a Soviet attack as well.

One cannot deny that a new tone has come to dominate Soviet leadership pronouncements on nuclear issues over the past decade. Yet the driving premises behind Soviet force development and operational planning remain disturbingly similar to those that have undergirded the Soviet buildup since its beginnings two decades ago. Whatever the more recent statements of the Soviet leadership may imply, ongoing trends in Soviet force modernization harbor operational consequences that are anything but congenial to Western security interests. What accounts for these apparent inconsistencies between Soviet rhetoric and Soviet behavior? It is to this complex issue that the present chapter is mainly addressed.

THE CASE OF THE TULA LINE

Let me first consider some of the changes in high-level Soviet declaratory commentary on nuclear policy that have taken place since the mid-1970s. Along with a parallel decline in the volume and specificity of doctrinal writing in the military press that has occurred during the same period, this shift has largely inspired those Western intimations of a "new" Soviet strategic outlook outlined earlier. The key question here concerns whether Soviet nuclear planning guidance has really evolved in consonance with the new leadership refrain or, conversely, whether the Soviet leaders have instead been manipulating foreign audiences with a studied propaganda campaign while proceeding with business as usual in the force development arena.

The roots of this issue lie in the emergence and progressive refinement of what has been called the "Tula line," so named after a major foreign policy address given by Brezhnev in that Soviet city in January 1977.[9] In that speech, the late Soviet leader endorsed a fundamentally new cast of Soviet declaratory policy that entailed some significant departures from previous doctrinal formulations. To begin, Brezhnev declared that the Soviet Union rejected strategic superiority as a force-posture goal and merely sought to be the military "equal" of the United States. More notably, he reversed two of the most enshrined axioms of Soviet military doctrine by disavowing any Soviet planning for preemption and claiming that no one could count on emerging from nuclear war the winner. Indeed, he went further yet by declaring that Soviet doctrine lacked any offensive content and was solely "defensive" in character. Finally, he laid the groundwork for what in 1982 became a formal theme in Soviet foreign propaganda —Moscow's accession to a unilateral nuclear no-first-use pledge.[10]

The combined message formed by these propositions soon became the predominant refrain of Soviet external commentary. Brezhnev routinely reiterated the main points of his Tula remarks at every opportunity and made a special point of emphasizing them in interviews with Western journalists. Soon, they began to crop up in the speeches of other Party figures and elsewhere in the Soviet media. Most interestingly, they began to be echoed by the Soviet military as well, first in a widely quoted *Soviet Military Encyclopedia* article by the former Chief of the General Staff, Marshal Ogarkov, and then in major pamphlets by Ogarkov and the late Minister of Defense, Marshal Ustinov.[11] In addition, this new declaratory refrain figured prominently in a Soviet Defense Ministry publication issued for external consumption in response to an earlier United States Defense Department report on the Soviet threat.[12]

In short order, the Tula line had become part and parcel of the Soviet military's declaratory posturing, expressed in terms indistinguishable from similar commentary emanating from the civilian side. In one of its more interesting variants, this new military pose came in the guise of retired Lieutenant General Milshtein, a former General Staff officer attached to Georgii Arbatov's USA and Canada Institute, repeatedly lecturing Americans during trips to the United

States about how ostensible "Red hawks," like Colonel Yevgenii Rybkin, should (depending on which version you heard) be dismissed as low-level functionaries who knew nothing of "real" Soviet defense policy or ignored as irrelevant military pedagogues who were not true professional officers.[13]

It would oversimplify things to discredit this new Soviet line out of hand, as some in the West have done, as nothing but cynical prevarication aimed at consumption by gullible outsiders.[14] For one thing, much of the Tula line can be reconciled with the classic principles of Soviet military thought, once the fine print on both sides is carefully scrutinized. For example, when Soviet leaders insist that nuclear war would be an unmitigated calamity, they are also careful to avoid adding any intimation that it is no longer a contingency the Soviet Union must continue to plan against. Granted, Brezhnev declared in 1981 that "it is dangerous madness to try to defeat each other in an arms race, to count on victory in a nuclear war."[15] But he did not say that victory was impossible. He only said that one should not count on it. Likewise, his statement that "only he who has decided to commit suicide can start a nuclear war in the hope of emerging victorious from it" fell considerably short of admitting that the Soviets had given up hedging their bets against the contingency all the same. Should war occur, Soviet doctrine (even at the political level) still counsels that Soviet involvement would be wholly legitimate and would aim toward advancing the triumph of socialism with every resource available. In effect, the Soviets appear to be seeking the best of two worlds: a combat capability whose credibility in the eyes of potential opponents might make war less likely and thereby enhance nuclear deterrence—but on terms congenial to parochial Soviet interests.

Beyond the compatibility of the Tula line with established principles of Soviet military doctrine, Western analysts have long recognized that the Soviets can scarcely lie to their own officers charged with implementing Soviet defense guidance merely in order to deceive foreigners. Although the Soviets have occasionally had to tap-dance smartly in their efforts to square the Tula line with the traditional injunctions of their military doctrine,[16] the language of the Tula refrain has generally been crafted so as to allow the Soviets to eat their cake and have it too. In fact, there is no incompatibility between Moscow's emphasis on war fighting in its defense preparations and its stress on deterrence at the state policy level. The issue has never been *whether* the Soviet leadership is committed to preventing nuclear war, but simply how they have chosen to accommodate that challenge. There is no reason why Soviet military leaders cannot easily live with political rhetoric dramatizing the notion that a nuclear war would be bad news for everybody, even as they prepare for precisely that eventuality.

Relatedly, when Soviet leaders insist that Soviet doctrine is not "offensive," they are referring to its political rather than its technical dimension. In effect, they are simply proclaiming that the Soviet Union has no intention of starting a nuclear war without provocation. There is nothing surprising about this proposition, and one must take care not to make more of it than it warrants. One need not posit an image of the USSR as a nation bent on war to demonstrate that its forces and employment concepts are configured to take the offensive whenever major war appears foreordained. In such a situation, of course, the Soviets would have to assess the risks of preemption against the costs of continued inaction in the face of uncertainty. Either way, their decision would be momentous.[17] Although there is no telling what counsel the Soviet leaders would follow if actually confronted with such a dilemma, there is little denying the course of action Soviet operational doctrine would prescribe—notwithstanding efforts by the Soviet media since the late 1970s to suggest otherwise.

Nevertheless, some troublesome aspects of the Tula line require further explication. Although outright disinformation may not be the whole story, there is clearly more than a trace of verbal sleight of hand involved. In engineering the Tula argument, the Soviets appear to have been quite successful in blending skillful mental reservation with a decided play on the political dimension of Soviet military doctrine in an effort to persuade outsiders that a major transformation has taken place in Soviet strategy.

For one thing, the phraseology that makes up the line is too neatly homogeneous to suggest anything other than careful orchestration. Indeed, the various signals emanating from Moscow on strategic matters since the Tula speech have come to reflect such consistency as to suggest that high-level guidance on approved language has been promulgated to all hands. Statements by senior military figures often read for all the world as though they were drafted by the same speechwriters who work for the Politburo. Furthermore, there has been an abrupt decline in significant Soviet public writing on military doctrine that appears too closely matched with the rise of the Tula line to be coincidental. Part of this, of course, may simply attest to a decline in open Party–military contention as the armed forces have grown increasingly satisfied with their resource allocations and institutional stature. All the same, the rich public discourse on nuclear matters that prevailed during the late 1960s and early 1970s has virtually vanished from the pages of *Krasnaia zvezda* and other military periodicals.[18] The few pieces on doctrine that do appear are little more than bland bromides that lace the most noncontroversial generalities about Soviet military thought with sprinklings of key phrases from the Tula line. This has certainly made life rougher for Western analysts of Soviet strategy, who must now wonder what sort of doctrinal rumination may be going on beneath public scrutiny. More important, however, it suggests that an indefinite moratorium may have been imposed on any open military commentary that does not show seemly obeisance to the current line.

My inclination is to regard the emergence of the Tula position and the concomitant demise of Soviet doctrinal writings as the results of mounting Soviet embarrassment over their own doctrinal hyperbole. After all, most Western analyses of Soviet military policy during the preceding decade reflected a decidedly conservative cast and drew their strength by quoting chapter and verse from the Soviet military literature. In light of this (perhaps bolstered by Moscow's determination to keep SALT on track in the face of progressively waning American enthusiasm for detente), the traditional Soviet doctrinal stress on the primacy of the offensive, on preemption and the feasibility of victory, and so on most likely became perceived by the leadership as a net liability. The Soviet response, in this interpretation at least, was to begin squelching loose talk in the military press in favor of emphasis on elements of Soviet doctrine that could be squared with preferred Western images of deterrence, parity, and stability.

In this connection, it was perhaps not accidental that within a year of the first articulation of the Tula line, the International Information Department of the CPSU Central Committee was formed under the leadership of Leonid Zamyatin and Valentin Falin, two accomplished public relations virtuosos. Although proof is hard to come by, numerous subsequent Soviet efforts to promote the Tula line have borne suspicious earmarks pointing toward this enterprising activity's involvement. These include, among other things, a rash of unusually slick Soviet pamphleteering against United States portrayals of the Soviet "threat";[19] the rare practice of allowing Soviet publicists to write letters and articles on nuclear issues in the Western media;[20] and the unprecedented use of high-level General Staff officers as front men to play up the Tula line in press conferences and audiences with foreign reporters. All in all, the thrust of rhetoric emanating from Moscow since the advent of the Tula position has had the effect of sounding like a casebook variant of Shakespeare's lady protesting too much.[21]

Several continuing undercurrents in Soviet policy make it particularly difficult to swallow Moscow's claims of late that its "old" doctrine has been invalidated by recent changes in the strategic environment. For one thing, concurrent writings in the military–technical domain (insofar as they have continued to address nuclear issues) have shown little deviation from the timeworn axioms of Soviet doctrine as far as operational matters are concerned. Second, military figures who have waxed so eloquently at one level about the centrality of the Tula principles in Soviet foreign policy have repeatedly given voice in other contexts to the continued primacy of all the Clausewitzian notions that have long made up traditional Soviet military strategy. Third, even Soviet pamphlets for external consumption that have expressly sought to promulgate the Tula message have exercised great care to avoid using formulations that would expressly contradict the classic teachings of Soviet military doctrine. Last, and most revealing, there have been no indications in observable Soviet research and development policy, force modernization, or day-to-day operations and training that suggest any fundamental departure from the basic policy guidance that has governed Soviet force development since the Soviet buildup first got under way two decades ago.

The Soviet no-first-use declaration likewise entails less than meets the eye at first glance. Although it has an impressive ring of magnanimity and lays a tough diplomatic challenge at the doorstep of the United States, it is not incompatible with Soviet military doctrine when examined in an operational context. Indeed, it makes sense on military as well as political grounds. Two circumstances make it cost free for the Soviets to embrace a no-first-use pledge. The first is the presence of parity in central systems, which has largely decoupled the United States intercontinental nuclear force from its erstwhile role in assuring escalation dominance for NATO in any European conflict. The other is the substantially improved Soviet conventional posture opposite NATO, which gives Soviet commanders a credible choice of conducting offensive operations without resorting to nuclear weapons, so long as NATO shows similar forbearance. Unlike the 1960s and early 1970s, when Soviet doctrine envisaged a massive combined-arms offensive in which nuclear strikes would occur conjointly with conventional operations, the Soviet armed forces today have a dual-capable posture that allows them the option of foregoing nuclear strikes at the outset of combat, should battlefield circumstances make this feasible. So long as their conventional offensive can be counted on to proceed according to expectations, they have no incentive to initiate nuclear operations and every reason to exercise restraint.

Indeed, there is logical ground to believe that Soviet commanders are now prepared to countenance a certain number of demonstrative nuclear strikes launched by NATO out of desperation without replying in kind, so long as those attacks remain symbolic and do not threaten militarily significant consequences. It is this capacity to accept at least a token NATO crossing of the nuclear threshold without blinking that enables the Soviets to get by with their unilateral no-first-use pledge. As Stephen Meyer has noted, Soviet leaders may be prepared to concede NATO the option of nuclear "first use" in the literal sense, but they have not yielded in their determination to land the first *decisive* nuclear blow in any situation in which they would risk defeat by doing otherwise.[22]

None of this is intended to suggest that Soviet doctrine is not amenable to change under the influence of evolving technology and strategic circumstances. It is only to argue that the various themes that make up the Tula line involve less a shift in Soviet doctrine per se than simply a new and imaginative departure in the particular emphases to which the leadership has chosen to play. When Soviet officials express such seemingly moderate notions as the importance of preventing nuclear war and stress the dire consequences that would befall any country that started one, they are merely underscoring the political dimension of Soviet doctrine, whose primary concern is deterrence. Likewise, when military figures reiterate the timeworn premises of Soviet doctrine regarding the imperatives of war waging, they are neither indicating resistance to the prevailing line nor clinging to outmoded axioms, but simply addressing the operational side of Soviet doctrine, whose primary concern is what to do when deterrence fails.[23] Both strains of rhetoric are part and parcel of a common strategic conception. If anything, their articulation in tandem attests mainly to an improvement in the adroitness of Soviet sloganeering. Any argument that they amount to more than that requires evidence that has yet to be presented.

SDI AND THE QUESTION OF HOMELAND DEFENSES

The Soviets have also sought to bring about such engineered revisionism in their recent stance on the ABM issue. In propagandizing against President Reagan's Strategic Defense Initiative (SDI), they have tried, among other things, to propound the idea that the Soviet Union has forsworn interest in strategic defense and has accepted mutual vulnerability as a suitable basis for Soviet security. As in the case of the Tula line, this claim is flatly contradicted by longstanding Soviet operational doctrine, as well as by amply funded Soviet programs in the Star Wars arena.[24] Nevertheless, Soviet commentary in the wake of President Reagan's March 1983 SDI speech has relentlessly hammered home the point that Moscow has changed its ways and now, in effect, embraces the logic of Western deterrence theory.

In this spirit, one commentator pointed out that although

accidental wars can always occur, under conditions of "genuine" stability (of the sort assertedly provided by the 1972 SALT accords), a nuclear war is "simply inconceivable."[25] The Soviets have also invoked the authority of their scientists (and frequently ours) in proclaiming that "no effective means of defense exist in nuclear war."[26] As for the incapacity of such defenses to protect large populations, Georgii Arbatov noted in passing how this "naïve concept" was shared by "maybe even some people on our side at the beginning"—implying that Soviet planners have since given it up.[27] His son Alexei, in a lengthy *SShA* article, added that although Soviet military thought had long embodied a deep strain of defense-mindedness, the Soviet leadership was not so foolish as to think it could fight and survive a nuclear war.[28] One could cite numerous variations on this theme. The point that matters, however, is that Moscow now pretends to accept the premises of mutual assured destruction (with implied willingness to settle for a deterrent grounded on Soviet vulnerability), but contends that Washington is seeking a nuclear disarming option through SDI, thus threatening to disrupt the balance that has hitherto prevented war.

This contention is virtually identical to the concern that has been widely voiced by defense officials of both the Reagan administration and its predecessors over the implications of Soviet programs that have been under way since the early 1970s. It is also refuted by tangible evidence of those programs, much of which the United States government has been reluctant to publicize for security reasons. Indeed, it is worth underscoring that the Soviet scientists and technicians who have been at the vanguard of Moscow's propaganda assault against SDI are the same individuals who have long figured at the center of the Soviet Union's own research and development efforts in ballistic missile defense (BMD).[29] Nevertheless, by putting Washington on the defensive with an argument that has a powerful appeal to those in the West inclined to believe it, the Soviet refrain against SDI has given Moscow another inside track in the propaganda war.

As for the argument often voiced in the West that Soviet BMD efforts reflect less than stellar technical promise, one must remember that 100-percent solutions have never been driving criteria behind Soviet defense planning. Soviet investment in homeland defense over the years has been consistent enough to suggest a Soviet determination to cope with the full spectrum of external threats despite the impossibility of total success, simply on the ground that failure to make the attempt would be politically irresponsible. Repeatedly, we have seen energetic (if less than effective) Soviet efforts to anticipate and offset every facet of the changing threat, ranging from bombers and cruise missiles to ballistic missiles and other space-related systems. In all cases, the abiding goals have been to preserve the "inviolability" of Soviet territory from "even one missile or one plane" and to repulse any attack so as to "ensure the uninterrupted work of the national economy, organs of state administration, and the combat capability of the armed forces."[30]

Obviously, a substantial discrepancy will remain, at least in the immediate years ahead, between this mission tasking and the Soviet Union's technical ability to make good on it. Yet this discouraging prospect shows no sign of being accepted as valid ground by Moscow to give up continued efforts all the same. Throughout the postwar era, the Soviet Union has sought to accommodate each changing element of the offensive threat, irrespective of the larger global political and diplomatic climate.[31] Even the ABM Treaty did not visibly shake Moscow's long-standing commitment to pressing the state of the art in strategic defense, whatever the current Soviet declaratory line may say to the contrary.

On this last point, it would leave the story unfinished not to place in context some recent Soviet intimations that the USSR has turned over a new leaf in its attitude toward ballistic missile defense. Georgii Arbatov, for example, ventured to sustain such a perception in Western eyes in his effort to deny that Moscow ever took seriously the promise of its ABM program, let alone was intimately involved in any such program today. Dismissing United States accusations to the contrary as "big inventions," Arbatov conceded in a prominent interview that "there was some work done" at one time, but insisted that it was "of very modest scope" and that the leadership "never had it in mind that it's possible to do it."[32]

Granted, the Soviet press has been conspicuously mute regarding BMD since the signing of the ABM Treaty, and one no longer reads vigorous expositions on homeland defense in any way resembling the doctrinal literature that existed on this issue more than a decade ago.[33] Nevertheless, Soviet research and development in missile defense have maintained a steady pace, as have the development and fielding of new technologies to counter United States airbreathing threats. Furthermore, there has not been the slightest indication that Moscow has repudiated the general mission of home defense in its strategic planning. One should not, of course, be surprised at the silencing of Soviet public commentary on BMD in the wake of the ABM Treaty's conclusion, in light of the regime's ample ability to maintain rigorous discipline over its internal media. Clearly, any such pronouncements would be impolitic in a situation in which the Soviets were attempting to dramatize their commitment to the spirit of the ABM Treaty.[34]

Nevertheless, the record belies the revisionist assertions of

Arbatov and others that Moscow has been unenthusiastic from the start about its involvement in BMD. On the contrary, one can muster a litany of authoritative Soviet statements from various times during the past two decades that ill fit the gospel according to Arbatov. On the eve of the Cuban crisis, for example, when the Soviet ABM effort was just beginning to produce its first primitive returns, Foreign Minister Gromyko lamented the persistence of the nuclear "balance of terror" and favored a new regime featuring a mutual builddown of offensive forces, yet with an "exception" for a "limited and agreed-to number of . . . antimissile missiles and antiaircraft missiles" intended "to guard against the eventuality, about which Western leaders have expressed anxiety, of someone deciding to violate the treaty and conceal missiles or combat aircraft."[35] A more widely cited expression of the same sentiment was voiced two years later by a prominent Soviet military theoretician, Major General Nikolai Talenskii, who wrote that "when the security of a state is based only on mutual deterrence by means of powerful nuclear missiles, it is directly dependent on the good will and designs of the other side, which is a highly uncertain premise." For that reason, Talenskii said, "it would hardly be in the interests of any peaceloving state to forgo the creation of its own effective means of defense against nuclear-missile aggression and make its security dependent only on deterrence, that is, on whether the other side will refrain from attacking."[36]

Lest one be tempted to dismiss Talenskii's remark as a parochial statement of military opinion not shared by the civilian rulers in the Kremlin, no less an authority than the late Premier Kosygin was moved to tell President Johnson at the Glassboro summit in 1967 that a ban on missile defenses was, in Henry Kissinger's words, "the most absurd proposition he had ever heard."[37] Kosygin gave voice to a similar outlook on the eve of the SALT talks in 1969 when he observed at a London press conference: "I think that a defense system which prevents [missile] attack is not a cause of the arms race. . . . Perhaps an antimissile system is more expensive than an offensive system, but its purpose is not to kill people but to save human lives."[38] Read out of context, that statement would strike most casual readers of the newspaper today as an exhortation by President Reagan on behalf of SDI.

To sum up, the weight of evidence regarding Soviet involvement in antimissile research, development, and testing, to say nothing of Moscow's apparent willingness to press to the edges of permissibility and beyond with respect to ABM Treaty compliance, casts serious doubt on the notion that the Soviets underwent "a change in their thinking about BMD" during the late 1960s.[39] Rather more persuasive, in my own view at least, is the proposition that "the Soviets consider strategic defenses neither inherently good nor inherently bad" and that their shift in declaratory rhetoric after the signing of the ABM Treaty "stemmed from Soviet assessments of the current strategic situation, not from categorical judgments about the inherent pros and cons of mutual national vulnerability."[40] As budgetary allocations to the Soviet home defense establishment in recent years (more than twice that devoted to United States *offensive* systems) bear powerful testimony, the Soviet military continues to assign high priority to missile and space defense, however deficient its prevailing technology and hardware base may remain.

CONCLUSIONS

To recall the subtitle of the conference for which this chapter was written ("Nuclear Strategy and Its Discontents"), there is no evidence I am aware of that the Soviet leaders are losing much sleep over problems with their nuclear strategy per se.[41] Insofar as they are afflicted by any "discontents" regarding the nuclear issue at all, those discontents are more likely to concern the unpleasant fact of American strategic competitiveness and the various obstacles it has placed in the path of making existing Soviet strategy work. Since the mid-1960s, Moscow has expended great efforts to bring Soviet forces into closer line with long-standing doctrinal edicts regarding the operational and technical prerequisites for coping with a breakdown of deterrence. The effect of this effort has been the gradual accretion by the Soviet Union of a plausible nuclear war-waging posture that might, in a reasonable Soviet interpretation, offer significant coercive advantages to the Soviet side in the event of a deep crisis or military collision between the superpowers. Seen in this light, the main consequence of the various United States force improvements that have been set in train under the so-called Reagan defense buildup has been to threaten a substantial erosion of Moscow's offensive advantages—and thereby to undermine the Soviet Union's evolving war option. Two rational choices that any Soviet leadership might entertain in the face of such a challenge would be (1) to seek solace in a new strategic paradigm or (2) to redouble their efforts at keeping their old one in force. Until now, at least as far as Soviet diplomatic and military conduct would suggest, the Kremlin's response has appeared to fall almost exclusively into the second category.

This chapter has emphasized the continuity of Soviet military thought during the past two decades regarding the most fundamental questions of war and peace, notwithstanding a pronounced shift since the late 1970s in the tone and emphases of Moscow's declaratory line on strategic matters. None of the foregoing, however, is intended to suggest that Soviet military concepts are unsusceptible to change at the operational level. On the contrary, the comprehensive improvements in Soviet nuclear and conventional forces since the advent of the Brezhnev regime not only have brought Soviet combat capabilities more and more into accord with Soviet doctrinal beliefs, as noted earlier, but also have made increasingly possible new Soviet force employment options either expressly ruled out or hitherto unaddressed by the Soviet literature.[42] In the years ahead, we can probably anticipate a continued maturation of Soviet doctrine, contingency plans, and operating repertoires in a way that accommodates the broadened range of options afforded by these new capabilities. Yet whatever changes that may occur are unlikely to make life for Western defense planners any easier. Although implementation strategies will continue to evolve in parallel with improvements in Soviet military technology, the underlying principles of Soviet doctrine, which stress the importance of assuring deterrence through the pursuit of plausible war options, show every likelihood of remaining established articles of faith.

As for the connection between Soviet doctrine and future Soviet responses to SDI, I would not ascribe a high probability to the Soviet Union's acceding to any agreement (whether through negotiation or tacitly) that seeks to replace the current nuclear relationship with a new arrangement dominated by homeland defenses on both sides. Quite apart from the fact that the Soviet Union has never shown much enthusiasm for subjecting its security to joint "legislation," as it were, by the involvement of the adversary, such a cooperative venture seems unlikely simply by virtue of Moscow's apparent preference to stick with more familiar solutions to its security problem.

Some officials in the United States government have suggested that President Reagan's much sought-after transition to a defense-dominated strategic world should be rendered that much easier by the "natural" preeminence of the home defense mission in Soviet military thought. My own sense is that this expectation stems from a fundamental misunderstanding of the role of strategic defense in Soviet military planning. Although Soviet military theory has long placed great weight on strategic defense, indeed far more than has been the case with American military policy, that emphasis has typically occurred within the context of a *continued parallel stress on the importance of overwhelming offensive forces.*[43] Whatever credence one may ascribe to the Soviet Union's putative doctrinal tradition of defensive-mindedness, the offensive remains the conceptual linchpin of Soviet strategy, and it is only through well-endowed offensive forces that Soviet planners contemplate winning any war they might have to fight in the wake of a catastrophic deterrence failure.[44] In this outlook, active defenses are merely envisaged as a backstop for what remains essentially a counterforce-oriented strategy. They certainly are not seen as a substitute for offensive forces or as suitable guarantors of Soviet security in and of themselves.

It is not, of course, inconceivable that the Soviets might, out of a common-sense desire to avoid worse alternatives, eventually recognize or be driven to accept the wisdom of joining in a cooperative effort with the United States to alter the existing foundations of the nuclear balance. But that will require their acceptance of a very different conceptual orientation than the one that currently energizes their ongoing force modernization. As matters stand, we may wait forever for Moscow to embrace the logic of defensive emphasis based on existing Soviet doctrinal proclivities.

NOTES

1. Bernard Brodie, *Escalation and the Nuclear Option* (Princeton, N.J.: Princeton University Press, 1966), p. 147.

2. Raymond L. Garthoff, *Soviet Strategy in the Nuclear Age* (New York: Frederick A. Praeger, 1958), p. xi.

3. "Nuclear Strategy: The Case for a Theory of Victory," *International Security*, Summer 1979, p. 6.

4. *Gulliver's Troubles, or the Setting of American Foreign Policy* (New York: McGraw-Hill, 1968), p. 160.

5. Robert S. McNamara, "The Military Role of Nuclear Weapons," *Foreign Affairs*, Fall 1983, pp. 65–66. Although it anticipates my argument to follow, I cannot resist noting here that this assessment comes from the same McNamara who, on the eve of the Soviet buildup in 1965, confidently intoned that "the Soviet leaders have decided that they have lost the quantitative race and . . . are not seeking to engage us in that contest. There is no indication that the Soviets are seeking to develop a nuclear force as large as ours." "Is Russia Slowing Down in Arms Race?" *U.S. News and World Report*, April 12, 1965, pp. 52–56.

6. David Holloway, *The Soviet Union and the Arms Race* (New Haven, Conn.: Yale University Press, 1984), p. 55. Unlike McNamara, however, Holloway does refer later to a "reformulation or clarification" of Soviet doctrine since the mid-1970s—which strikes me as coming much closer to getting the point right—and recognizes the persistence of a Soviet belief that nuclear war "has to be prepared for" and that "strategy . . . has not been made redundant" (p. 167).

7. Dimitri Simes, "Deterrence and Coercion in Soviet Policy," *International Security*, Winter 1980/81, p. 88. Like Holloway, Simes also qualifies his assessment by noting that the Soviets have had "sound reasons of public

relations to restrain their militaristic rhetoric" (p. 90). I will return to this point in some detail later.

8. Raymond L. Garthoff, "Mutual Deterrence, Parity, and Strategic Arms Limitation in Soviet Policy," in *Soviet Military Thinking*, ed. Derek Leebaert (London: George Allen and Unwin, 1981), pp. 93, 104. This argument is further developed in Garthoff's more recent *Detente and Confrontation: American–Soviet Relations from Nixon to Reagan* (Washington, D.C.: The Brookings Institution, 1985), pp. 768–85.

9. "Speech of Comrade L. I. Brezhnev," *Izvestiia*, January 19, 1977. To the best of my knowledge, the term was first suggested by Andrew Goldberg, "The Tula Line: Change and Continuity in Soviet Nuclear Doctrine Since 1977," unpublished manuscript, October 1983.

10. TASS communique, June 15, 1982.

11. See, respectively, Marshal N. Ogarkov, "Military Strategy," in *Soviet Military Encyclopedia*, Vol. 7 (Moscow: Voenizdat, 1979), pp. 555–65; Ogarkov, *Always in Readiness to Defend the Homeland* (Moscow: Voenizdat, 1982); and Marshal of the Soviet Union D. Ustinov, *We Serve the Homeland and the Cause of Communism* (Moscow: Voenizdat, 1982).

12. *Whence the Threat to Peace* (Moscow: Military Publishing House, USSR Ministry of Defense, 1982, in English). Among other things, this booklet dismissed as "nonexistent" those Soviet writings and statements cited by Western analysts as evidence that "Soviet military doctrine is of an aggressively offensive nature and that the Soviet Union counts on winning a nuclear war by means of a preemptive strike" (p. 11). The use of language here is clever. No responsible analyst that I know of has suggested either that Soviet doctrine is "aggressively" offensive or that the Soviet Union "counts on" achieving victory through nuclear preemption.

13. Seminar at the Rand Corporation, Santa Monica, California, October 1, 1979. In a later interview, Milshtein went even further in trying to distance the current Soviet line from earlier military writings: "To arrive at a correct evaluation of our doctrine today, you must not rely on Sokolovskii's book but on those quite unambiguous Soviet documents and statements that have appeared in more recent years." Anthony Austin, "Moscow Expert Says U.S. Errs on Soviet War Aims," *New York Times*, August 25, 1980. Even the most case-hardened Soviet military ideologues, however, recognize the dangers of nuclear war and reject the idea that it can be considered an acceptable instrument of Soviet policy. For example, the most renowned of the so-called Red hawks, Colonel Ye. Rybkin, wrote in 1973 that "nuclear weapons will cause very serious destruction and an unprecedented number of victims. . . . Recognition of the fact that a war involving the use of nuclear weapons remains a just war on our part does not at all mean that we welcome such a war. . . . A nuclear war on the part of socialism can only be a *forced* continuation of politics and a retaliatory defensive step against the aggressors." *War and Politics in the Modern Era* (Moscow: Voenizdat, 1973), p. 27. Eight years earlier, Rybkin had authored a hard-hitting piece whose most arresting passage argued that "any a priori rejection of the possibility of victory is harmful because it leads to moral disarmament, to a disbelief in victory, and to fatalism and passivity." "On the Essence of a World Nuclear-Missile War,"

Kommunist vooruzhenykh sil, No. 17, September 1965, pp. 50–56. For a translation and analysis, see Roman Kolkowicz, *The Red "Hawks" on the Rationality of Nuclear War* (Santa Monica: The Rand Corporation, RM–4899–PR, March 1966).

14. See, for example, Joseph D. Douglass, "Soviet Disinformation," *Strategic Review*, Winter 1981, pp. 19–21.

15. *Pravda*, October 21, 1981.

16. For example, in a 1981 article Marshall Ogarkov showed due obeisance to Brezhnev's stock refrain that "our strategic doctrine has a strictly defensive orientation." Yet he promptly added that "it also envisages, in the event of an attack by an aggressor . . . resolute actions by the Soviet armed forces, which have full mastery of the art of waging not only defensive, but also modern offensive operations on land, in the air, and at sea." Marshal N. Ogarkov, "Guarding Peaceful Labor," *Kommunist*, No. 2, February 1981, p. 86. A similar indication of military discomfort over the "defensive emphasis" theme was apparent in the following invocation of Lenin by a prominent Soviet military theoretician: "To tell us that we must wage only a defensive war when the knife continues to be raised above us . . . is to repeat old phrases of petty bourgeois pacifism which long ago lost their meaning." Colonel General N. Lomov and Colonel S. Alferov, "On the Question of Soviet Military Doctrine," *Voenno-istoricheskii zhurnal*, No. 7, 1978, p. 25.

17. For elaboration, see my "Uncertainties for the Soviet War Planner," *International Security*, Winter 1982/83, pp. 139–66.

18. Some of the more significant of these writings can be found in translation in William R. Kintner and Harriet Fast Scott, eds., *The Nuclear Revolution in Soviet Military Affairs* (Norman: University of Oklahoma Press, 1968).

19. In addition to *Whence the Threat to Peace* noted earlier, see also *The Threat to Europe* (Moscow: Progress Publishers, 1981). The latter document repeats all the choicest quotables from Brezhnev, Ustinov, and Ogarkov and shrugs off the more disturbing themes of the Soviet military literature as the product of "military theorists devoted not to doctrine or military policy but to particular aspects of combat" that "necessarily envisages the training of soldiers for various actions on the field of battle" (p. 10). Yet this is precisely the point. Soviet statements that disavow "aggressive" intent or "counting on" nuclear victory refer exclusively to *premeditated* nuclear war—which no serious Western analyst has accused the Soviets of planning to begin with. They say nothing, however, that would deny Soviet planning (and associated force development) for massive preemption in any crisis in which the Soviets were convinced that deterrence was about to fail.

20. Among many examples that could be cited, see Radomir Bogdanov and Lev Semeiko, "Soviet Military Might: A Soviet View," *Fortune*, February 26, 1979, pp. 46–48; Henry Trofimenko, *Changing Attitudes toward Deterrence*, ACIS Working Paper No. 25 (Los Angeles: Center for International and Strategic Affairs, UCLA, July 1980); Nikolai Chervov, "A Soviet View on Nuclear War," Letter to the Editor, *Los Angeles Times*, February 12, 1982; and Vitaly Zhurkin, "A Soviet View: MX Is a First-Strike Weapon," *San Jose Mercury News*, May 1, 1983.

21. This post-Tula turn of Soviet declaratory posturing on the nuclear issue reminds one of Strobe Talbott's comment several years ago about how "there has never been anything more offensive than a Russian on the defensive." "Whatever Happened to Detente?" *Time*, June 23, 1980, p. 34.

22. There have been occasional intimations in Soviet doctrinal writing that in some circumstances Soviet commanders might forgo responding with nuclear weapons to an isolated enemy battlefield nuclear strike. There are also indications that this option has figured in Soviet training exercises. In all events, as Meyer has pointed out, "it is not the first nuclear use per se that is of concern to Soviet military planners, so much as the first decisive use of nuclear weapons in the theater." *Soviet Theater Nuclear Forces, Part I: Development of Doctrine and Objectives*, Adelphi Papers No. 187 (London: International Institute for Strategic Studies, 1983), p. 28.

23. Such imputation of a split between so-called unilateralists and diplomacists in the Soviet defense debate is the only unpersuasive part of an otherwise excellent treatment of the Tula issue by Dan L. Strode and Rebecca V. Strode, "Diplomacy and Defense in Soviet National Security Policy," *International Security*, Fall 1983, pp. 91–116.

24. For further discussion, see Benjamin S. Lambeth, *The Soviet Union and the Strategic Defense Initiative: Preliminary Findings and Impressions*, The Rand Corporation, N–2482–AF, June 1986.

25. L. Semeiko, "Stability or Destabilization? Two Approaches to the Path of Development in the Contemporary World," *Izvestiia*, June 14, 1984.

26. "Space Deceivers," *Argumenty i fakty*, September 20, 1983.

27. Radio Moscow international service, April 13, 1983.

28. A. G. Arbatov, "Limiting Antimissile Systems: Problems, Lessons, and Prospects," *SShA: ekonomika, politika, ideologiia*, No. 12, December 1984, pp. 16–28. Arbatov junior qualified for first prize several months later in grasping for a novel argument to invoke against SDI. In response to a query from a panel of West German interviewers as to why the USSR was so resolutely opposed to Star Wars if, as Soviet propaganda alleged, it not only was technically infeasible but also could drive the American economy into the ground, Arbatov deadpanned that this was undesirable from the Kremlin's point of view, since Moscow knew that any such "economic crisis" would impose intolerable burdens on the American "proletariat," who would be "first in line" to feel its effects! "Auch Inspektionen vor Ort sind Moglich," *Der Spiegel*, March 11, 1985.

29. For example, Academician Yevgenii Velikhov wrote in an opinion piece in the *Washington Post* on June 24, 1983, that strategic defense is "a dream that can't come true." Yet he runs the Kurchatov Atomic Energy Institute and has been identified by the CIA as a "central figure" in Soviet laser and particle-beam weapons research. The same is true of Nikolai Basov, another prominent Soviet anti-SDI propagandist, who was a 1964 Nobel laureate in quantum electronics and has evidently spent much of his professional life working on both conventional and exotic BMD technologies. See Lord Chalfont, "Moscow's Star Wars Plan: Keeping Facts under Wraps," *Toronto*

Globe and Mail, April 23, 1985; and Bill Gertz, "CIA: Soviets Are Developing Their Own 'Star Wars' System," *Washington Times*, May 10, 1985.

30. Colonel N. Tabunov, "In the Spirit of Personal Responsibility," *Vestnik protivovozdushnoi oborony*, June 1975, p. 11; Lieutenant General N. Grishkov, "To Strengthen the Might of Air Defense," *Vestnik protivovozdushnoi oborony*, August 1972, p. 2.

31. This development pattern has stood at notable odds with Andrew Marshall's otherwise apt observation that the strategic interaction process has been "muffled, lagged, and very complex" in its hardware manifestations. Quoted in Graham T. Allison, *Essence of Decision: Explaining the Cuban Missile Crisis* (Boston: Little, Brown and Company, 1971), p. 98.

32. "Star Wars Will Ruin All Arms Control Negotiations," interview in *U.S. News and World Report*, September 30, 1985, p. 27.

33. In perhaps an unwitting reflection of the "left hand knoweth not" syndrome in Soviet politics, however, there appeared in 1977 a fascinating account of an obscure Soviet novel that romanticized the struggle for technical excellence, amidst all varieties of bureaucratic subterfuge, in a fictional effort to develop and field the USSR's "first antimissile missile complex." This instructive morality tale, which really warrants a separate analysis all to itself, lent persuasive support to the notion that whatever the agreed line on BMD may be in Soviet *external* discourse, the subject of ABM remains very much a live one within internal Soviet military and defense–industrial circles. See the review by Engineer Colonel General (Ret.) N. M. Popov of the book *Bitva* ("The Battle") by Nikolai Gorbachev in *Knizhnoye obozreniye* (Moscow), No. 46, November 18, 1977, pp. 8–10.

34. Interestingly, this imposition of a gag rule did not entirely squelch internal Soviet military comment on BMD-related subjects. In a review of pertinent Soviet materials published since the conclusion of the ABM Treaty, the head of the Policy Planning Staff in the FRG Defense Ministry recently determined that "although the Soviet literature refrained from any direct reference to the significance of missile defense, some writers showed that it was still possible to allude to the subject by extending the scope of air defense to space, as it were, and even by referring on occasion in this connection to the requirements of defense against 'ballistic' attack." Hans Ruhle, "Gorbachev's 'Star Wars,' " *NATO Review*, August 1985, p. 29.

35. Speech to the U.N. General Assembly, September 26, 1962, excerpted in "When the Soviets Liked Superpower Defenses," *Wall Street Journal*, January 15, 1985.

36. "Antimissile Systems and Disarmament," reprinted in *The Military–Technical Revolution*, ed. John Erickson (New York: Frederick A. Praeger, 1966), pp. 225–7. In a direct contradiction of Talenskii's statement, two prominent Soviet Star Wars critics more recently contended that "it is impossible to ensure reliably the security of a state by means of military–technical innovations. Even the very illusion of the possible achievement of such security is extraordinarily dangerous." Ye. Velikhov and A. Kokoshin, "Nuclear Weapons and the Dilemmas of International Security," *Mirovaia ekonomika i mezhdunarodnye otnosheniia*, No. 4, April 1985.

37. Henry Kissinger, *White House Years* (Boston: Little, Brown and Company, 1980), p. 208.

38. Quoted in Ernest Conine, "Soviets Ride a Crooked Mile in Space," *Los Angeles Times*, August 26, 1985.

39. David Holloway, "The Strategic Defense Initiative and the Soviet Union," *Daedalus*, Summer 1985, p. 259.

40. David B. Rivkin, Jr., "What Does Moscow Think?" *Foreign Policy*, Summer 1985, pp. 93–95.

41. Indeed, it would come closer to the mark, I believe, to say that they have drawn substantial political advantage from Western manifestations of such discomfiture, whether these have come in the form of handwringing over the specter of "nuclear winter," *cris de coeur* over the immorality of deterrence, lamenting the failure of the United States to accede to a nuclear freeze, or whatever.

42. Examples include limited strategic warfare; counterforce-only attacks; selective attacks against enemy command and control facilities; war in space; crisis management and escalation control; and the use of land-based missiles against targets at sea.

43. For example, at the time the United States was just beginning its pre–SALT I exploratory probings on the possibility of an ABM freeze, a Soviet military writer entered a strong plea for continued Soviet attention to offensive force enhancement by stressing the primacy of the offensive as the only way to "achieve the routing of the enemy and establish victory" in modern war. Colonel S. Krupnov, "On the Development of the Methods and Forms of Armed Conflict," *Krasnaia zvezda*, January 7, 1966. A similar view was expressed more recently in Marshal Ogarkov's injunction that although Soviet "strategic doctrine has a strictly defensive orientation, it also envisages, in the event of an attack by an aggressor . . . resolute actions by the Soviet armed forces, which have full mastery of the art of waging not only defensive, but also modern offensive operations on land, in the air, and at sea." Marshal N. Ogarkov, "Guarding Peaceful Labor," *Kommunist*, No. 2, 1981, p. 86.

44. For an early Soviet statement that defense is not an end in itself but rather is integrally tied (through its damage-limiting function) to the overall offensive mission, see Major General I. Zavialov, "On Soviet Military Doctrine," *Krasnaia zvezda*, March 30, 1967. This view has been reaffirmed in more recent Soviet military writing: "The main form of military action will be the strategic offensive. As to the strategic defensive, it is only acceptable as a temporary condition for holding off the offensive of enemy strategic groupings." Lieutenant General M. Kir'yan, *Military–Technical Progress and the Armed Forces of the USSR* (Moscow: Voenizdat, 1982), p. 315.

PART III

The Future of Deterrence

CHAPTER 8

The Long-Term Future of Deterrence

JOSEPH S. NYE, JR.

Can nuclear deterrence last forever? To answer yes requires heroic assumptions. Simple but powerful intuitions about human fallibility and skepticism about complex systems may account for the greater pessimism about deterrence on the part of the general public than on the part of nuclear strategists. President Reagan's speech of March 1983 may have been technologically premature, but it was timely in terms of domestic politics. According to some accounts, it was a response to the public mood, as well as to some of his military advisers' views that current policies offered little hope of relief in the long term.[1]

Generally speaking, nuclear strategists have ignored the question of the long-term future of deterrence. The rational models of deterrence theory tend to be static and ahistorical. However, when a president announces not merely a hope but also a massive program to make nuclear weapons "impotent and obsolete" and asks "How long, after all, can the American people hold to a strategy that threatens innocent lives?" strategists cannot continue to duck the question.[2] Many analysts complain about the alleged lack of technical or political realism in the president's proposal, but few have sketched alternative visions or even seriously addressed the problem of how to think about the long term.

The Strategic Defense Initiative (SDI) is only one among a number of relevant utopias that can be compared in terms of their intrinsic desirability, their political and technical conditions for stability, and the relative feasibility of processes for their attainment. It is time that strategic studies move out beyond static models and address more adequately questions about the long-term future of deterrence. Such a task is more easily said than done. This chapter

Portions of this chapter draw on my book, *Nuclear Ethics* (New York: Free Press, 1986).

merely tries to clarify the problem and suggest a research agenda, rather than to provide a set of answers.

EVIDENCE, PROBABILITY, AND METAPHORS

Deterrence seems to have worked for forty years. Pro-deterrence strategists make too much and the antinuclear critics make too little of the absence of any war between the superpowers for the past four decades. The simple form of the pronuclear argument begs the difficult causal question of ascertaining what would have been the case in the absence of nuclear deterrence, and more important, it also begs the question of why the future should resemble the past. For those who believe that catastrophic failure is inevitable, it is no answer to say simply that deterrence has worked in the past.

At the same time, the argument that deterrence has worked is not without merit. Four decades without war among the great powers is a remarkable period of peace in modern Western history. (The record is forty-three years between 1871 and 1914.) There is good reason to believe that nuclear weapons contributed to the prudence that kept leaders out of war during the past forty years.[3] Nuclear weapons may have had something like a crystal ball effect. Imagine what would have happened if the statesmen who led the world into this century's first great conflagration in 1914 had possessed a crystal ball showing them the world of 1918. The leaders of 1914 expected a short, sharp war followed by business as usual. One suspects that if the German Kaiser, the Russian Czar, and the Austrian Emperor had seen a picture of 1918—with their thrones vacant, their empires destroyed—they would have drawn back from the brink of war that summer. Today, modern leaders know the horrible devastation that would result from any nuclear war. In the words of the Summit communique, "recognizing that any conflict between the U.S.S.R. and the U.S. could have catastrophic consequences, they emphasized the importance of preventing any war between them, whether nuclear or conventional."[4]

This is only partial comfort, however, because crystal balls can be shattered by accident or miscalculation. Even if nuclear deterrence has lasted for nearly four decades, it is difficult to believe it will last forever. Some dire fears, such as C. P. Snow's 1960 prediction that nuclear war within a decade was a mathematical certainty, have proved wrong. It has been argued that Snow's exaggeration will be justified if nuclear war takes place in the next hundred years.[5] But the difference between a decade and a century can make all the difference in the world when it comes to deciding what can be done to avoid nuclear war.

Some antinuclear critics object to the whole approach to deterrence in terms of probability. For example, the sociologist Todd Gitlin argues that "since deterrence works only if it works forever, it is an all-or-nothing proposition, so applying the language of probability to it is misleading."[6] But his argument is not compelling. Gitlin seems to assume that failures of deterrence or inevitable accidents must lead to all-out nuclear war, but this is far from self-evident. Indeed, a case can be made that an accident or a partial failure of deterrence may be the prelude to major changes that reduce risks or reliance on nuclear deterrence in the long term. Even if he were right about catastrophe, however, it is odd for Gitlin to discount "microscopic probability" by asking, "Do we feel secure playing Russian roulette if the revolver has a hundred chambers?" Perhaps not, but if he had to play, I doubt that he really would not care whether a revolver had one hundred chambers rather than six! And if he readmits probabilistic reasoning, then it can also be applied to the question of relative risks between unilateral disarmament ("refusing to play") and trying to increase the number of chambers in a world in which the game of nuclear deterrence already exists.

Other antinuclear critics use the concept of probability to "prove" that deterrence will inevitably fail. With a series of trials over time, even a low probability approaches a certainty. The probability of at least one successful event in a series of trials is equal to $1 - (1 - p)^n$. Thus, if we flip a coin once, the chances of getting tails once is 50 percent. If we flip a (true) coin ten times, the chance of seeing tails at least once is 99.9 percent. Using this approach to the distribution of probabilities, Bradford Lyttle concludes that the average daily chances of nuclear war during the past thirty years (10,950 days) was probably not higher than .0006 percent. But Lyttle goes on to argue that even if the daily chances of one missile's being launched are only one in a hundred million, the probability of such an event passes 50 percent within fifty years.[7] Douglas Lackey points out that a one-in-a-hundred chance of nuclear war in the next forty years becomes a 99 percent probability after eight thousand years.

Although such calculations can be useful numerical reminders, we cannot conclude very much about deterrence from them. They assume that probabilities are constant and that events are independent of each other. The metaphor of a flipped coin is misleading. Human interactions are more like loaded dice. The odds change; and the outcome of one set of events may greatly change the odds of the next event. In fact, frightening events like the Berlin and Cuban missile crises may drive the odds of war down in their immediate aftermath. A diplomatic rapprochement, such as occurred between the United States and China in the 1970s, can have a similar effect. New technologies may cut either way. It seems unlikely that the odds of nuclear war future generations face will be the same as ours.

Alternatively, some antinuclear critics who believe in the inevitability of nuclear war make the opposite assumption—that the odds are not independent but are driven in a perverse direction by the interaction of events.[8] But unless this is an example of the gambler's fallacy that a series of events is evidence for an increased probability of more such events (that fellow is on a roll), there must be a specified causal model to make such an argument plausible. Such causal models can be evaluated in the light of evidence. Often, like the propositions that arms races inevitably lead to war or that we have been piling up more and more destructive capacity, the models turn out to be weak on the facts.[9]

The likelihood of nuclear war rests on both independent and interdependent probabilities that relate different aspects of the process by which war might occur. Purely accidental war might be conceived of in terms of independent probabilities; but if the numbers are low enough, they may not matter. If we are speaking of eight thousand years, for example, humankind may have concerns other than nuclear war. The longer the period, the greater the chance that other things will have changed, and that an accident will have a different meaning in that changed context. In short, even when we have independent probabilities (as we might approximate in the case of pure accidents), the significance of the events they lead to must be seen in the context of the interdependent probabilities involved in human history. As Paul Schroeder has put it, "Murphy's Law does not apply to history." That is not an argument for optimism, but it is "an argument *against* a certain extremely popular kind of crippling pessimism."[10] Those who use simple probabilistic arguments to argue that abolition of nuclear weapons is the only policy that is both moral and realistic build their case on shaky grounds.

A more sophisticated antinuclear argument against probabilistic reasoning is presented by the British philosopher, Robert Goodin. He argues that all notions of probability and likelihood are simply inappropriate "because we just do not know enough about the shape of the underlying distribution to justify employing any of the standard techniques for estimating probabilities. . . . Thirty-five years is just too short a run." What is more, we lack well-validated scientific theories about nuclear war from which we might derive probabilities, and psychological evidence shows that individuals are poor probability assessors. Thus, nuclear deterrence is reckless "playing the odds without knowing the odds." Instead of probabilities, we should use a "possibilistic" logic with only three crude categories—certain, impossible, and possible—to assess consequences.[11]

Goodin's points are a useful caution against spurious precision in estimates of probabilities that strategists often make. He warns us against the efforts to "fine-tune" deterrence and justify huge expenditures on new weapons systems by tiny increments in deterrence. But Goodin's arguments are less convincing as a reason to lump all probabilities into one crude category of "equally possible." When we deal with rare events like nuclear wars, where there can be no series of tests, probability is bound to be subjective. But that does not make the concept useless. The chance that a certain football team will win a championship may be one in a million, yet it is possible to find a bookie who will take a bet. "In such a circumstance, that is exactly what is meant by probability, the odds that a rational and informed person will bet on the occurrence of the event. . . . It is, in fact, an informed guess."[12]

Our knowledge about nuclear deterrence is limited, but it is not totally absent. We have evidence from forty years of experience. Even if theory is weak in international relations, it is not totally absent. The evidence from psychological experiments that Goodin quotes shows that subjective estimates of probability are often wrong. It also shows that estimates by specialists tend to be more reliable than estimates by the general public. Moreover, the public tends to overestimate unfamiliar risks and underestimate familiar ones.[13]

What are the risks of nuclear war? This is a critical question, but unfortunately there is no fully satisfactory answer. In recent years, public opinion polls have shown from one-third to one-half of the American public saying they expected nuclear war with the Soviet Union within the next few years. In contrast, my informal polling of a few score of nuclear specialists during the past few years has produced a modal answer of expectations of nuclear war between the United States and the Soviet Union in the next decade of about one chance in one hundred. The range of answers among experts, however, stretched from one in five to one in a million. This compares with a known probability of being killed in an automobile accident in the same period of about one chance in five hundred.

The point is not to prove that the public or the specialists are wrong, but merely to remind us of the range of uncertainty in any estimates of the risks of nuclear war. Despite our inability to be precise, we are bound to make crude estimates of whether the probability is high or low. It seems more plausible to assume that the probabilities are relatively low than that they are very high. There are both rational and accidental models of the causes of war, and both models seem to have low probabilities in the nuclear age. As McGeorge Bundy has written, "In light of the certain prospect of retaliation, there has been literally no chance at all that any sane political authority, in either the United States or in the Soviet Union, would consciously choose to start a nuclear war."[14] As for pure accident, the prospects seem low. In Michael Howard's view, it is hard to find any historical evidence of a purely accidental war. Technical progress during the past two decades has reduced the prospects of purely accidental onset of nuclear war.[15] In other words, there are good theoretical reasons to believe that the experience of the last forty years was not implausible evidence that the unknown probability is unlikely to be at the high end of the distribution of probabilities. We cannot say that the annual average probability cannot be higher than one in forty, but we can say, on the basis of the distribution of probabilities, that it is unlikely that it was very much higher than that. The 95 percent confidence level would be .072. Of course, average probabilities can be misleading if trends or unexpected events cause changes, but they provide a rough starting point or base line. Chances of nuclear war in the range of 2 to 7 percent are far from C. P. Snow's "certainty," although they are still uncomfortably high! We should make neither too much nor too little of the forty years we have experienced.

Unfortunately, it is also difficult to be precise about trends in probability over time. For example, is the current situation more or less risky than the period of previous concern, that is, 1958–1962? Those who argue that nuclear risk was higher in 1962 point to such technical improvements as permissive action links; improved command, control, and communication; national technical means of verification; and such political factors as United States and Soviet experience in managing crises. Those who argue that the risk is higher in the current period point to the loss of United States nuclear superiority, the greater Soviet capability to support forces in third world areas, the deployment of vulnerable weapons and support systems that place a premium on preemption, doctrinal stress on protracted war fighting, and the deterioration of political dialogue.

In examining such arguments, it is interesting to note the mixture of technical and political factors, and the mixture of arguments that rest on assumptions about rational actors and those that stress nonrational and accidental factors. Policy responses vary accordingly.

Within a rational actor framework, the insanity of large-scale nuclear war (i.e., the extreme disproportion between political ends sought and the consequences of the military means used) suggests that nuclear war is very unlikely. At the same time, political conflict will occur, and we will need to deter a variety of Soviet actions. Hawks argue that the danger of nuclear war comes from a Soviet miscalculation of the credibility of capability of our commitment. The way to reduce risk and to enhance deterrence is to add nuclear capability so that it is clear we cannot be beaten at the end of a multiple move game (or escalation ladder).

The rational actor model can also lead to other conclusions. Many doves tend to believe that provocation is a greater danger than is temptation in the onset of war, and that the hawks' unrestricted armament could create such provocation. Although the historical evidence that arms races lead to war is more shaky than prevalent views admit, there are historical cases, most prominently American policy toward Japan in 1941, in which one country intended to deter another, but succeeded only in provoking the second country to initiate war.[16] The proper response in such cases is reassurance, although hawks and doves differ on the relative dangers of provocation versus appeasement and on the proper balance to be struck between deterrence and reassurance.[17]

A third view of the probability of the onset of nuclear war rests on a nonrational model that includes such factors as psychic stress, misperception, bureaucratic pathologies, and accidents. Owls worry about loss of control more than wrong intention. Although pure accidents under normal conditions are extremely unlikely, the probability of accident and inadvertent initiation can rise dramatically in a situation of nuclear crisis. At such times, the safety catches that make accidents unlikely are deliberately released, psychic stress is increased, and there is little time to correct mistakes. Such crises also present the rare situation in which, on an expected value definition of rationality, it might make sense to initiate war. If one were absolutely convinced that the other side was about to launch a nuclear attack, it might seem better to strike first rather than second.

In short, the interaction of nonrational and rational factors in crisis situations that might involve escalation and preemption seem the most likely causes of dramatic rises in the probability of nuclear war. Although all three models of the potential onset of nuclear war capture important aspects of causation, many strategists have focused too much on policies of deterrence and reassurance related to intentions, and too little on situations in which intentions may be irrelevant or so constrained that rational choice becomes virtually meaningless.

TECHNOLOGICAL AND POLITICAL UTOPIAS

It does not follow from the proposition that the system of deterrence might fail someday that nuclear holocaust is either imminent or inevitable. Nonetheless, any chance of holocaust or species extinction is so horrible that we would be wise to reduce our reliance on nuclear weapons as much as possible in a way that does not raise the current risks of nuclear war. Given the fallibility of human reason and human organizations, and the inevitable spread of nuclear technologies to more countries (and possibly terrorist groups), it would seem wise gradually to reduce our reliance on nuclear weapons. Whether this can ever lead to abolition of weapons is impossible to say. Certainly, it would be impossible to abolish nuclear knowledge without burning all books and all scientists. The prospect for that solution may have passed when the Pope failed to burn Galileo! However, even if the knowledge of nuclear weapons is an inescapable part of the trajectory of human history, there is nothing inevitable about the number of weapons and the centrality of their role in military and political relations among nations.

Robert Tucker argues that "the first, and indeed the last thing that needs to be said in any moral assessment of deterrence, is that it should not be regarded as a temporary arrangement. . . . Its justification now cannot be made to turn on a prospect that from the vantage point of the present must appear as near utopian. . . . Deterrence, it appears, will remain a part of our 'condition' for as far as we can presently see."[18] Tucker is correct to caution us against overly optimistic attempts to escape deterrence in the near term; such utopian thinking could have highly immoral, unintended consequences. But he would be wrong if he were to discourage long-term utopian thinking. We simply know too little about the distant future to make such categorical assertions, and a little dose of utopian thinking now about the distant future may involve few short-term risks, while helping us to set our policies in a direction in which currently improbable solutions become possible some time in the future. Almost by definition, the current probabilities of utopian solutions are low, but they may become higher for future generations. We simply do not know. In any case, we should avoid foreclosing such prospects and take proportionate risks to keep such possibilities open.

How might one envisage moving toward less reliance on nuclear weapons and lower risks of nuclear war in the future? Two centuries ago, Immanuel Kant suggested that peace among nations might come about as a result of three things: the increasing destructiveness of war; the spread of republican governments; and the growth of commerce and trade. In the short run of two centuries, Kant proved wrong. Extensive trade and enormous destructive capability did not prevent Europe from destroying itself in World War I. But perhaps a century and a half was too short to test the validity of Kant's prescriptions. Unlike 1914, when many leaders (and their citizens) glorified and desired war, nuclear devastation has made war far less attractive today. Although the evidence is not conclusive, it is almost impossible to find instances of war in this century among republican (i.e., constitutional liberal representative) governments.[19]

Trying to predict long-term change would be foolish, although one can sketch rough outlines of broad paths of change that might lead to a future with less reliance on nuclear weapons. Roughly speaking, there are technological paths and political paths. The latter are the more promising, but the former are now receiving more attention.

Among the most dramatic technological changes of the past decade has been the increase in the accuracy with which weapons can be delivered. This accuracy range has reduced the nuclear destructive power needed to destroy particular targets and has increased the precision with which some military targets can be separated from civilian ones. In addition, there have been important improvements in the identification of changing targets and the flexibility of the weapons that might attack them. These changes cut two ways. On one hand, they permit greater discrimination and the possibility of counter combatant targeting; on the other, they give rise to the temptation to think of nuclear weapons as militarily useful in battle. From an owlish perspective, this concept would be seriously mistaken. But the same improvements in accuracy and identification of targets can be used to improve the capabilities of conventional weapons and allow them to replace battlefield nuclear weapons. In other words, if wisely applied, the technology of accuracy could help reduce reliance on battlefield nuclear weapons and on city-burning strategic weapons. However, technology alone will not determine that outcome.

The technological promise that has received greatest attention in recent years is the prospect of effective defense against ballistic missiles. President Reagan expressed a desire to escape from the dilemmas of nuclear deterrence in his speech of March 1983. However, escape from deterrence requires a leak-proof defense, not only against ballistic missiles but also against bombers and cruise missiles. Such a perfect defense seems unlikely. Some who doubt the technological feasibility of such a task have urged that the new Strategic Defense Initiative Organization in the Pentagon concentrate on the lesser task of defending missiles rather than cities. But this is a rationale that would enhance rather than replace deterrence.

Moreover, the feasibility of a utopia is more than a matter of technology. It also has a political dimension. Even if the complex defense technology can be developed and combined into a system that has to be perfect without ever having been tested under the stress of nuclear war, can we get from here to there without going through a transition period in which the nuclear predicament of both sides would be worse? Will improving defenses stimulate a massive increase in offensive weapons as the other side tries to prove it can still overcome the defenses? If it appears we are about to perfect a defensive system in a few years that would effectively disarm the Soviet deterrent, would they take more risks during crises in the short run because they saw the likelihood that these risks would pay off politically becoming bleaker and bleaker with the passage of time?

These questions do not mean that there should not be a significant research program on strategic defense. Quite the contrary. The prospect of enhancing deterrence, and perhaps, in the long term, of being able to save a large number of lives in case deterrence fails, justifies considerable efforts. The value of such an initiative will depend on the consequences, not the motives. It may be "better to defend than to avenge," but only if the consequences of *trying* to defend do not increase the risk of nuclear conflict in the meantime. Those consequences are likely to be determined not by our intentions, but by the types of technology chosen (how it will affect crisis stability) and the political state of United States–Soviet relations (will the introduction of defense be in a cooperative or an antagonist setting?).

The political path lacks the glamour of the technological one, and as many technologists have discovered, politics is harder than physics. But political change is the key factor in the long-range future. Further, political change may take a number of forms. There may be changes in relations between the United States and the Soviet Union; there may be changes in the growth of international institutions and cooperation among states; there may be changes in domestic political and social attitudes toward the sovereign state and its defenses.

Sometimes political relations between states change quickly —witness the change in relations between the United States and China in less than a decade. Because such rapid changes are often related to the existence of a common enemy, it would be unwise to expect a similarly rapid change in United States–Soviet relations.

There are several deep-seated reasons to expect tension in United States–Soviet relations. First, as de Tocqueville already saw in the nineteenth century, the enormous size and resources of the Russian and American nations foreshadowed a future bipolar rivalry. Then in 1917, the Bolshevik Revolution added a layer of deep ideological incompatibility. When the Second World War destroyed the multination balance of power existing before 1939, it left a bipolar structure of world power centered on the United States–Soviet rivalry. The accumulation of vast nuclear arsenals overshadowing those of all other nations has consolidated that special relationship. From a power politics point of view, the probability of tension is built into the very structure of the relationship.

Nonetheless, at different times, there have been lesser degrees of hostility. It is worth remembering that the worst outcome has not occurred. Despite hostility, there has also been prudence in managing the world's first nuclear balance of power. The destructiveness of nuclear weapons introduced a disproportion between most ends that the superpowers seek and the principal military means at their disposal.

This situation has led to the evolution, by a process of trial and error, of some primitive rules for avoiding or managing crises. The rules are so primitive that they might more correctly be called prudent practices. Indeed, they began well before the onset of detente in the 1970s and survived its demise. As described by Stanley Hoffmann, "One such informal rule was the non-resort to atomic weapons. A second rule was the avoidance of direct military clashes between armed forces. A third element was the slow (and for America) painful learning of limited wars . . . calculated so as to limit the risks of escalation, even if those constraints made a clearcut victory or a rapid settlement impossible. Later came the beginnings of nuclear arms control between Washington and Moscow."[20] Rudimentary as these rules or prudent practices are, they are significant if one believes that a stable balance of power requires a degree of moderation in the actors' behavior, as well as a military balance.

When there is a degree of common interest in stability, the balance of power can become a positive sum game in which both sides win. Indeed, an international *regime*—a set of tacit or explicit rules and procedures—may be developed to encourage a stabilizing perspective of long- rather than short-range self-interest. The nineteenth-century balance of power and the contrast between Bismarck's restraint and the failure of his successors is often cited as an example.

To say that the postwar United States–Soviet balance of power has been embedded in such a regime would be to stretch the point. The rules are ambiguous and not openly accepted by both sides. In 1972 and 1973, Nixon and Brezhnev signed agreements that seemed to codify the rules, but the ambiguities (such as exceptions for "wars of national liberation") later led to a sense of cheating and deception after the Middle East War of 1973 and the Soviet transport of Cuban troops to Angola in 1975 and 1976. Moreover, an agreed regime implies reciprocity and flexibility in bargaining behavior as the opponents seek to avoid jeopardizing the regime. In United States–Soviet relations, however, reciprocity has proved to be limited in time and on issues. It is difficult to "bank goodwill" from one time or issue to another in the relationship. Whereas both sides have a degree of interest in preserving stability in the form of preserving nuclear bipolarity and the avoidance of nuclear war, their competition makes it impossible to agree on the status quo. While a common interest exists, it is severely limited and strained by the competitive dimension.

Theorists of international cooperation point out, however, that warm relations and trust are not necessary for cooperation.[21] Competition does not prevent cooperation, and a central authority is not necessary. What is needed is a more sophisticated and long-range perception of self-interest, an ability to learn from experience, and the realization that a relationship will continue over time. When games of "Prisoners Dilemma" are played over long periods, a strategy of reciprocity proves to be most effective. On such realist premises, it is quite possible to expect a gradual evolution of United States–Soviet cooperation. Certainly, relations in the 1980s are vastly different from those in the 1950s. Despite frequent allegations of return to the Cold War, there is much more contact and cooperation in the 1980s than there was thirty years ago.[22] There is far greater awareness of each other's perception of its vital interests and of its internal processes. It is highly plausible to expect change in the relationship over the next thirty years and more.

In addition, there may be changes inside the Soviet Union. The Soviet Union is unlikely to change quickly, but it does evolve slowly and unevenly. It has opened up somewhat during the last thirty years. There are more contacts as well as more sophistication in the perception of outside reality. There are more pinholes letting light into the "black box."

Beyond the United States–Soviet relationship, there may be other political changes that will affect the role of nuclear weapons in world politics. That the development of cooperation among states rests on self-interest rather than goodwill is encouraging because it implies that cooperation is consistent with realistic premises. There may continue to be a gradual development of international regimes that govern various dimensions of economic and social interdependence among states. In some cases, these regimes may involve a gradual growth of institutions. In other instances, economic integration among states may reduce the degree of potential conflict, as it has between ancient enemies such as France and Germany. The development of transnational institutions and contacts may gradually transform domestic attitudes toward sovereignty and the use of violence to defend the state. After all, the modern territorial state that replaced looser feudal loyalties has only been the dominant institution since the Peace of Westphalia in 1648. It is not implausible to think it may give way to other institutional forms, if we look forward for an equivalent period.[23]

Of course, this Kantian view of the distant future is only one among a vast number of possible outcomes. Other benign (or horrendous) outcomes are also possible. The important point is that we should not let our imaginations be captured solely by images of imminent nuclear holocaust or by cynical views of the immutability of our dependence on nuclear weapons, for both images tend to stifle the modest efforts we can and should make now so that a Kantian type of future will be somewhat more probable for coming generations to develop. Similarly, we should not let the current debate about a single technological utopia like SDI so dominate our thinking that we fail to evaluate its various dimensions in relation to alternative visions of the long-term future.

ASSESSING END POINT AND PROCESS UTOPIAS

It is not enough merely to sketch the existence of potential benign alternative visions of the long-term future. Nearly everyone can think of seven impossible things before breakfast on any given day. We also need a framework for assessing alternative visions.

Most utopian visions tend to posit some better set of conditions or end points in the future. Those end points can be described in narrow military or broad political terms. Such end points can be compared in terms of their intrinsic desirability. For example, a global hegemonic dictatorship might reduce nuclear threats to the survival of the species, but at an enormous cost in terms of liberties. End point utopias can also be compared in terms of their political and technological conditions for stability. History will not stand still when a better end point is first reached. What are the conditions required to make it stable? For example, both a world of few nuclear weapons and a world of near perfect defense may require similar technical conditions of invulnerability (for the few remaining weapons or for the defensive systems) combined with a high degree of political trust and cooperation to be stable.[24] Third, one can compare the relative difficulty of achieving a smooth transition to the various hypothesized better end points. For example, those who believe in the virtues of multipolarity can argue that it is almost certainly easier to reach a multipolar balance of power (for example, by encouraging Japan to develop nuclear capabilities) than to establish hegemonic control or world government.

Table 8.1 illustrates an approach to the comparison of various end point utopias specified in terms of the role of nuclear weapons.[25] Each end point utopia rests on different assumptions about the rational and nonrational causes of war (Model 1 and Model 2); each rests on different political and technological conditions for stability (including conventional deterrence); and each can be assessed in terms

Table 8.1 *Assessing nuclear end point utopias*

Nuclear Weapons	*Intrinsic Desirability*	*Political and Technological Conditions for Stability M_1 and M_2*	*Relative Political and Technological Feasibility*
1. Accurate low yield weapons only			
2. Unlimited invulnerable weapons			
3. Finite deterrence			
4. Very low numbers			
5. Abolition (non-nuclear deterrence)			
6. Active defense			
7. Lengthen time before a decision for use can be made			

of relative feasibility. Those with greater intrinsic desirability, and those that require less heroic assumptions, appear to be the most promising for further exploration.

Table 8.2 presents an analogous approach in terms of international political organization, ranging from world government through various configurations of interstate power to situations of weakened nation-states located in a dense web of transnational and subnational institutions and solidarities, which might be called a *new feudalism*. Combinations and military and institutional end points can also be compared.

It would be a mistake, however, to limit our investigation solely to the comparison of end point utopias. One should also consider process utopias. A process utopia can be defined as a hypothesized benign or pacific trend, though the end point of the trend is uncertain. If one could be certain that a process was consistently reducing nuclear risk over time, that could be a more powerful source of optimism than the positing of a particular but infeasible end point. For example, Richard Garwin argues that if the current probability of nuclear war were one in one hundred, and we could each year "reduce the probability to only 80 percent of what it was the previous year, then the cumulative probability of nuclear war for all time will be 5 percent."[26]

Thus, as we look at process utopias, we want to know about the clarity of the hypothesized trend, the conditions that affect the

Table 8.2 *Assessing international institutional end point utopias*

Organization	*Intrinsic Desirability*	*Conditions for Stability*	*Relative Feasibility*
1. World empire			
2. World government			
3. Hegemonic control			
4. Superpower condominium			
5. Collective security regimes			
6. Multiple balance			
7. Regional spheres			
8. Proliferated world			
9. New feudalism (non-state-centered royalties)			

strength of its effects, and the prospect for being able to affect the process or encourage it in a benign direction. One can find a variety of process utopias both in historic "peace literature" and in social science theory. Such processes can be clustered under the political, social, and technological headings.

It would be a mistake, however, to focus solely on benign processes. Social science theory also provides a number of hypotheses about malign trends, even if their end points are uncertain. There are nightmares as well as dreams, and the same questions about clarity of the trend, conditions that affect the strength of effects, and the prospects for identifying manipulable variables can be asked about process dystopias as well as process utopias (see Table 8.3).

CONCLUSIONS

Strategists have paid too little attention to the long-term future of deterrence. In part, this may be a result of the static nature of deterrence theory; in part, it reflects the vague and unwieldy nature of the subject. The field was left to the antinuclear peace activists. The wishful or apocalyptic nature of their formulations led to easy dismissal and reinforced the strategists' natural tendency to avoid the subject. However, the general public has grown more concerned about the long-term future of deterrence, and most recently, President Reagan preempted the issue with his SDI proposal.

Thus, strategists are being drawn to a subject they were re-

Table 8.3 *Assessing process utopias and nightmares (hypothesized benign or malign directions although the end point is not certain)*

Processes	*Clarity of Trend*	*Desirability of Effects*	*Feasible End Points*	*Manipulable Variables*
1. Political				
2. Social				
3. Technological				

Notes:

1. Political Processes
   a. Cooperation theory: evolution of cooperation among states. Regimes for crisis prevention. How does theory fit past United States–Soviet behavior? Causation? Substitution? Policy implications?
   b. Integration theory: transformation of hostile relationships and growth of pluralistic security communities. Past cases? Effects of trade and social interdependence? Implications for United States–Soviet relations?
   c. Political learning: elite and mass perceptions? Small-group processes? Bureaucracies? Generational? Role of crisis? Role of ideology? The development of de facto doctrines of nonuse of weapons?
   d. Conflict resolution theories: applicability to international level?
   e. Nightmares—hegemonic transition theory: is bipolarity eroding? Soviety or U.S. decline? How perceived? How affects behavior? Manageable?

2. Social Processes
   a. Evolution of national and transnational norms: theories of multiple loyalties and control of aggression. World opinion. Experience with slavery, colonialism. Rates of diffusion.
   b. Evolution of Soviet and American societies and foreign policy. Diminished role of ideology; increased risk aversion; generational change; contact and perception.
   c. Economic determinism. Theories of modernization, rationality, and democratization producing pacific behavior.
   d. Nightmares—growth of revolutionary or millenial movements with access to technology of mass destruction. What potential threats? What access? What motivation for use? What effects?

3. Technological Processes
   a. Reducing probability of inadvertent war through changes in types and controls of weapons.
   b. Reducing incentives for attack through changes in types and controls of weapons.
   c. Nightmares—trends toward more vulnerable, less controllable, more accident-prone systems. What are potential trends? How dangerous? How manageable?

luctant to address. Many of their early efforts in the Star Wars debate have shown a rather narrow and sterile approach. In a sense, this approach reflects the sterility of deterrence theory as an academic subject. The major work was done two decades ago by applying powerful rational actor models (borrowed from game theory and microeconomics) to a confused politics of nuclear threat and counterthreat. Although there have been some useful efforts to enrich the impoverished psychology and lack of institutional context of the original models, the field has grown more in the quantity of work than in the quality of theory. Now, the pressure of events may lead strategists to reexamine the relationship of deterrence theory to its wider political context. If so, the field may be the better for it.

In any case, politicians and the public will not wait for academic theorists. Political actors of all stripes are proposing a variety of end point and process utopias. It would be a pity if academic analysts let the accidents of current events and political debate prevent a proper comparative analysis of these utopias. SDI and abolition are merely two among a number of contenders. This chapter does not propose an answer to the issue of the long-term future of deterrence, but it does suggest a framework for clearer thinking and an approach to research on the question.

NOTES

1. See Robert Scheer, "Scientists Scramble to Fulfill Plan," *Los Angeles Times*, September 22, 1985.

2. Quoted in *New York Times*, October 16, 1985.

3. John Lewis Gaddis, "The Long Peace," *International Security*, Vol. 10 (Spring 1986).

4. *New York Times*, November 22, 1985.

5. Thomas Powers, *Thinking about the Next War* (New York: Knopf, 1982), p. 17.

6. Todd Gitlin, "Time to Move beyond Deterrence," *The Nation*, December 22, 1984, p. 676.

7. Bradford Lyttle, *The Flaw in Deterrence* (Chicago: Midwest Pacifist Publishing Center, 1983), pp. 8–11. Note that Lyttle's numbers depend on the number of missiles that could be accidentally launched. Nuclear war need not follow.

8. I am indebted to James Blight for this point and for helpful conversations on this section in general.

9. See "Arms Races and the Causes of War, 1850–1945," in Paul Kennedy, *Strategy and Diplomacy* (London: George Allen and Unwin, 1983); and Samuel P. Huntington, "Arms Races: Prerequisites and Results," *Public Policy*, 1958, pp. 41–83.

10. Paul Schroeder, "Does Murphy's Law Apply to History?" *The Wilson Quarterly*, Winter 1985, p. 88.

11. Robert E. Goodin, "Nuclear Disarmament as a Moral Certainty," *Ethics*, March 1985.

12. H. W. Lewis, "Technological Risk," unpublished paper.

13. For example, Paul Slovic, Baruch Fischoff, and Sarah Lichtenstein, "Facts and Fears: Understanding Perceived Risk," in *Societal Risk Assessment*, ed. Richard Schwing and Walter Abers (New York: Plenum Press, 1980).

14. McGeorge Bundy, "To Cap the Volcano," *Foreign Affairs*, Vol. 47, 1969.

15. Michael Howard, *The Causes of Wars* (London: Unwin Paperbacks, 1983); and Paul Bracken, "Accidental Nuclear War," in *Hawks, Doves and Owls*, ed. Graham Allison, Albert Carnesale, and Joseph Nye (New York: Norton, 1985).

16. Graham Allison, Albert Carnesale, and Joseph S. Nye, eds. *Hawks, Doves and Owls* (New York: Norton, 1985), chapter 8.

17. See Michael McGuire, "The Dilemmas and Delusions of Deterrence," in *The Nuclear Crisis Reader*, ed. Gwyn Prins (New York: Vintage, 1984), p. 78.

18. Robert W. Tucker, "Morality and Deterrence," *Ethics*, April 1985.

19. See Michael Doyle, "Kant, Liberal Legacies and Foreign Affairs" *Philosophy and Public Affairs*, Vol. 12, nos. 3 and 4.

20. Stanley Hoffmann, *Primacy of World Order* (New York: McGraw Hill, 1981), p. 11.

21. See Robert Keohane, *Beyond Hegemony* (Princeton, N.J.: Princeton University Press, 1984); and Robert Axelrod, *The Evolution of Cooperation* (New York: Basic Books, 1984).

22. See Joseph S. Nye, Jr., ed. *The Making of America's Soviet Policy* (New Haven, Conn.: Yale University Press, 1984).

23. See Joseph S. Nye, Jr., *Peace in Parts* (Boston: Little, Brown, 1971) for a discussion of federalist functionalist and neo-functionalist strategies of building peace. See also Karl W. Deutsch et al., *Political Community in the North Atlantic Area* (Princeton, N.J.: Princeton University Press, 1957) for a discussion of "pluralistic security communities" where states have reasonable expectations of nonviolent behavior.

24. See, for example, Charles L. Glaser, "Why Even Good Defenses May Be Bad," *International Security*, Vol. 9, Fall, 1984.

25. I am indebted to James Blight and other members of the Avoiding Nuclear War Project at the Harvard Center for Science and International Affairs for discussions of this section.

26. Quoted in "Is There a Way Out?" *Harpers*, June 1985, p. 39. These issues are discussed further in J. S. Nye, Jr., *Nuclear Ethics* (New York: Free Press, 1986).

CHAPTER 9

Nuclear Deterrence and "World War III": How Will History Judge?

KEN BOOTH

For forty years, we have walked through the valley of the shadow of nuclear deterrence. And we have threatened and been threatened by great evil. There has been order between the major power blocs, but also unprecedented danger. Many of us believe the risks of breakdown have grown in recent years, albeit slightly: certainly the fears of nearly all of us have grown regarding the nature of the catastrophe that would result. Even if the risks of a breakdown are not as great as many of us believe, what might be destroyed is so infinitely precious that even a slight increase in the possibility of nuclear war can never be other than of compelling importance.

STRATEGIC THINKING IN DISARRAY

Thinking about strategy is at a premium, but it is more uncertain now than it has been for a long time. The established framework of theories and policies about peace and security in the nuclear age has come under increasing challenge. This is a time for judgment. It is sometimes said that an age passes when its basic illusions crack, and Hegel said that an understanding of a social system comes only when it is on the road to extinction. Since the late 1970s, the golden age of nuclear deterrence, spawned by the ideas and technology of the cold war, has seemed to be cracking. And in its gathering dusk, Minerva's owl has begun its flight. Strategic history, in the Hegelian manner, should lead the wise person and drag the fool; but who is the wise person, and in which direction will he or she be led?

We all have some understanding of the destructive power created by the nuclear–technological revolution, but it is by no means clear, judging by our words and actions, that we all fully grasp its enormity. Ever since the Great War, in Paul Fussell's words, "The drift of modern history domesticates the fantastic and normalises the unspeakable."[1] The devastation of Hiroshima and Nagasaki give only the merest hint of the nuclear catastrophe now possible: yet, many people have learned to live with the bomb. Growing numbers have not, however, and they have helped to stimulate the ferment on questions of security that we have seen since the late 1970s.

How are we to grasp our predicament? We realize that we cannot disinvent nuclear weapons; but we know that they can all too quickly disinvent our fragile social, economic, and cultural structures, the products of centuries. In these circumstances, we can only rush to judgment and have the "courage of our confusions." The instant destructive capability of the nuclear age demands instant historical judgment. If history can be stopped short by nuclear weapons, it is only right, before it is too late, that the condemned have a say.

The theory and practice of nuclear strategy in the West has been in crisis for several years (that it may not be in the Soviet Union says more about the anachronistic state of that country than it does about the state of nuclear strategy). If we are to restore the consensus and develop more successful theories and policies in future, it is necessary to generate more accurate images of the strategic world, for the reality we seek to cope with is partly defined by our manner of conceiving it. Strategic reality, like beauty, is in the eye of the beholder, and what passes for strategic realism these days is too often based on doctrinaire assumptions rather than on sophisticated appraisal of the facts.[2] Strategic realism is frequently out of touch with world politics. This can be seen in what history will judge to have been three of the major features of contemporary nuclear strategy: (1) the pervasive semantic debasement that characterizes the way we talk about the subject; (2) the nuclear hypochondria that dominates its theory and practice; and (3) the tunnel vision that characterizes risk assessment and the definition of what constitutes our real problem.

SEMANTICS OF NUCLEAR STRATEGY

People who love words make unhappy strategists. There always have been word problems in contemporary strategy: some people have sought to impress by the use of jargon; some have used pseudoscientific language to try to make the subject what it is not; and others have sought to hide its brutal side behind a veil of euphemisms.

A more fundamental problem now results from the fact that our basic concepts have become anachronistic: they have lost touch with the world they seek to analyze and influence. While this is so, the strategic world that is discussed is more the product of the ideology of Realism than of the realism that comes from empirical analysis. Good strategy must therefore begin with the defense of the language.

The words at the very heart of our subject—*war*, *strategy*, and *weapon*—decisively change their meaning when the adjective *nuclear* is attached to them. Even so, we continue to use these words, for the most part without a second thought, though the technology of mass destruction had deprived them of their earlier connotations with reasonable instrumentality. We try to handle the apocalyptic problems of today (and the future) with the Clausewitzian concepts of the past. But even such words as *apocalyptic*, *disaster*, and *catastrophe* are too frivolous to describe the events that would follow general war.[3]

In the extreme case, nuclear technology has destroyed any reasonable relationship between political ends and military means. Our collective strategic imagination is constrained by our words, but for the moment we remain gripped by the Clausewitzian language of reasonable instrumentality. It would be desirable to scrap obsolescent concepts, like obsolescent technology, but it seems to be impossible. War, strategy, and weapon are too deeply embedded in the way we think about the game of nations. As a result, all that can be done is mentally to place inverted commas around war, strategy, and weapon whenever they are preceded by nuclear. This should help remind us that the preeminent strategic–semantic fact of our time is that nuclear war, nuclear strategy, and nuclear weapon are contradictions in terms.

In light of the above, we can immediately define World War III out of our future. Unfortunately, this does not mean that there is no prospect of an all-out clash of violence between the nuclear powers. It only means that such a conflict, in the nuclear age, would not belong to a series that began with World Wars I and II: an all-out nuclear conflagration would be altogether different from such military antiquities. Consequently, those who profess to think the unthinkable by writing future histories about World War III or by advocating that nuclear war be conceived in traditional terms delude themselves and endanger everybody. Such an event would be a nightmare not a war, and to the extent that such works have become popular and influential, they have been pernicious.[4] Many nuclear strategists criticize the peace movement's dwelling on nuclear horror, and they claim to be no less sensitive to what is involved: but if this is the case, why do some of them write as if they have no imagination? Those who think in terms of World War III show the same blindness as those strategists in the past who sought to fight the next war with the concepts of the last. We are no better as strategists than our forefathers. Indeed, in some ways we are worse. Excuses can be found to justify the faulty assessments of war before 1914 and 1939: we can have no excuses. We have witnessed Hiroshima, Nagasaki, Bikini, and the rest. We should know better. But old concepts, like old soldiers, never die, and until they fade away we will continue to connive in socializing into international politics the ultimate horror threatening civilization.

Many of our major strategic concepts disintegrate on touch. They cannot therefore be the foundation for a science of strategy, or an art. We cannot hope to ease our problems until we have adequate words for contemporary realities. The phrase *conventional war*, for example, has all the wrong connotations when applied to modern weapons and vulnerable industrial societies. And what of the distinction between tactical and strategic, when tactical nuclear weapons can lay waste a thousand years of European civilization? It is truly a Humpty-Dumpty world when strategists can talk, apparently seriously, of war fighting in the nuclear context. Would it be war (or hell)? Would it be fighting (or massacre)? The phrase *war fighting* shows the extent to which contemporary strategy has to borrow Fussell's phrase, "normalised the unspeakable."

Language that is anachronistic must, perforce, produce anachronistic thinking. This is shown by the nuclear winter hypothesis, which makes clear that in at least some conceivable circumstances even our fears about nuclear war have been understated. Now, at last, it seems that we really can wage a war to end all wars. Whether or not those qualified to enter the scientific debate decide in time that nuclear winter is a probability or merely a possibility, its inherent doomsday potential should change our attitudes, for nuclear strategy has within it the power to be the final solution to the problem of international politics. Given the world-shattering risks of escalation, nuclear weapons cannot be used rationally. If they cannot be used rationally, can they sensibly be threatened? And if they cannot sensibly be threatened, what then is this thing called *nuclear deterrence*?

THE IDEOLOGY OF NUCLEAR DETERRENCE

For three decades, nuclear deterrence was the jewel in the crown of strategic sophistication. Until the mid-1970s, most strategists were dazzled by its brilliance. It promised to solve our major security problems, simply and cheaply. As a result, some theorists exhibited a Trotsky-like enthusiasm and naïvety: they seemed to believe it would merely be necessary to issue a few deterrence proclamations and then shut up shop. Not for the first time, the complexities of international politics scorned those who offered simple remedies. Since the mid-1970s, deterrence as a concept and a policy has come under increasing strain: it is now one of the several shopworn remnants of the golden age of contemporary strategy.[5] Its critics have increased in number; those who once thought they understood it are now less certain; and even its strongest supporters are not as complacent as they were. Though no generally acceptable alternative seems readily at hand, nuclear deterrence is facing more than an "artificial crisis".[6] Critics from the right and the left of the political spectrum have been pointing out its weaknesses and dangers and have been arguing, for different reasons, that its time is limited. How we move on will be crucial, for in some scenarios there is no civilized life after deterrence.

Nuclear deterrence has attracted criticism from a variety of standpoints. It is charged, inter alia, with being oversimplified, overblown, apolitical, immoral, illogical, too technological, too rational, and too irrational.[7] Such criticisms now constitute a formidable body of charges, and there is little value in rehearsing them. Instead, this chapter discusses that dimension of the problem sometimes called the *ideology of deterrence*.[8] This is a more fundamental problem and obstacle to change than anything else, for particular weapons and policies can always be altered more readily than men's minds.

Nuclear deterrence has become the opium of the defense people. Originally taken up to cope with the trials of the cold war, it has subsequently become a major problem in its own right. Nuclear deterrence ceased simply to be a strategic idea; instead, it took over the minds of many of its supporters.

An ideology is a program of action deriving from a particular view of the world. An ideology analyzes the world for its adherents; it explains what should move them, and to what goals; it offers guides to conduct; it justifies and legitimizes actions; it offers rationalizations for behavior; it socializes its supporters and provides a common language to bind them together; it consoles the true believers on the chosen road; and it establishes limits or boundaries on permissible thought and action. Nuclear deterrence has become an ideology in this sense. It constitutes a cluster of interrelated beliefs about the deployment and employment of nuclear weapons in international politics; it offers analysis, explanations, goals, guides, justifications, consolations, limitations on thought, and group bonding. The detailed character of these beliefs will vary according to different national strategic milieus, but in the West they have been very closely associated with the bundle of theories that grew out of the golden age of nuclear strategy between the mid-1950s and mid-1960s.

Ideologies can be addictive, and nuclear deterrence has proved to be so. The interplay of theory and practice, the cumulative change from one act of threat to the next, and the sense of success that has been engendered, have led to "aggravation."[9] What started out as a rational, indeed natural, reaction to the circumstances of the time has been pushed too far. What was grasped as the solution to the East–West conflict in the early days of the cold war has now become an important part of the problem. As a result, we now have a strategic community of compulsive nuclear deterrers: they cannot have enough of deterrence, whatever the state of international relations; deterrence is a habit of thought that they cannot break; deterrence, with all its repetitive and trivial routines, consoles them; and its litany curbs speculation outside the established framework. When deterrence becomes compulsive or addictive, it can lead to nuclear hypochondria and semantic debasement, a dulling of the critical and moral faculties, an illusion of success and rightness, irrationality (a reluctance to consider alternative approaches), an intolerance of criticism, the manifestations of "groupthink," the confusion of ends and means and a loss of perspective about what constitutes the real problem. When an ideology becomes an addiction, especially in the business of strategy, it can be dangerous to the health not only of its adherents but also to everybody else.

THE STRATEGIC FUNDAMENTALISTS

Before examining some effects of deterrence ideology, it is necessary to identify its chief prophets. (The criticisms advanced in this chapter are not directed at everyone who believes in nuclear deterrence, nor are they directed at the idea of deterrence. Such a position would be foolish. Who does not believe that nuclear weapons usually induce caution in the minds of decision makers? And who in the defense debate does not believe in deterrence in some form or another? Nonviolent civilian resistance, after all, is a form of deterrence.) The target of this essay is that group of American strategic fundamentalists whose voices have increasingly dominated the nuclear debate since the mid-1970s, and who, through the Reagan administration, have had some opportunity to put their ideas into practice. The street demonstrations of the antinuclear left may have animated public consciousness in the West, but the agenda in professional and policy circles has come under the sway of the pronuclear new right.

Characteristic attitudes and prescriptions advanced by the strategic fundamentalists of the new right are as follows:

1. The United States government should not just talk about a nuclear war-fighting capability, but should exert its efforts and provide itself with one.
2. United States doctrine should be based on the assumption that nuclear war can and should be controlled.
3. War planning can make a difference between which side prevails and survives passably well, even after a nuclear war, and this should inform the making of doctrine.
4. Nuclear war, though a calamity, is not the worst catastrophe conceivable.
5. The character of Soviet behavior greatly depends on the state of the military balance, so that any weakness shown by the United States will tempt expansion or exploitation.
6. The United States should establish whatever military edge it can over the Soviet Union in the nuclear field.
7. The Soviet Union is an evil empire that lies and cheats and only understands force.
8. United States security cannot be reliably based on bargains with the Soviet Union such as those implied in mutual assured destruction and arms control.
9. Detente in the 1970s was an illusion that the Soviet Union sought to exploit in order to increase its power and cover its expansionism.
10. Relations with the Soviet Union are a zero sum game.
11. United States strategists are in the business of increasing the strength of their own nation, and this is the best guarantee of international security.
12. Arms control frequently hobbles sensible military planning, while undercutting public support for defense, and so it has only benefited the Soviet Union.
13. The only useful role for arms control is to manipulate public opinion by exploiting its potential for propaganda.
14. The objective of the United States should not be to live with the Soviet Union but to act in such a way as to bring about the dissolution of the external Soviet empire and a radical transformation internally.
15. If the worst occurs and nuclear war breaks out, the United States has the right to seek victory, even though the world might perish in the process.

The twin pillars on which the prescriptions of the strategic fundamentalists rest, therefore, are a commitment to nuclear war fighting and a deep anti-Sovietism.

The views just described have been and are put forward with varying degrees of sophistication. They have been crudely advanced by unthinking hawks. More important, they have been developed and propagated by clever, articulate, and knowledgeable individuals —those who might be dubbed the "best and rightest" in American politics.[10] This latter group, abrasive in its approach and influential in its efforts, has at its core the members of the Committee on the Present Danger. They were the vehicle in the second half of the 1970s for the dissemination of neo-rightist fundamentalism regarding United States strategy and foreign policy. In the 1980s, many of the individuals associated with the Committee moved into important positions in the Reagan administration. Although some of the remarks of this group, both inside and outside of government, can be dismissed as rhetoric to impress friends and enemies alike, many represent fundamental beliefs. It is not likely that these beliefs have changed, even if their public presentation has been somewhat revised.

The domination of the defense debate in the West by these strategic fundamentalists has had many practical and theoretical implications. These, and criticisms of them, are discussed through the course of this essay. Confronted by the weight and typically combative style of the new right, the traditional deterrers stepped into the wings. In addition, President Reagan skillfully managed to identify criticism as being unpatriotic. In the mid-1980s, there has been something of a fight back by the traditional deterrers, but the important debate about relations with the Soviet Union has become sidetracked, like so much else, by the Star Wars gimmick. Unfortunately, the task of wresting the agenda from the strategic fundamentalists will not be easy, for their influence now goes deep in United States policy-making circles, while their fundamentalist counterparts in the Soviet Union play into their hands. Like security, insecurity is largely a reciprocal process: the more United States strategy is dominated by the deeply anti-Soviet counterforce advocates, the stronger will be the position of the strategic fundamentalists in Moscow. And so it goes. In any event, given Soviet "strategic culture," with its emphasis on deterrence through defense, its xenophobia, its faith in numbers, its traditionally crude power politics approach to the game of nations, and its belief that security comes through independent strength, the task of devising methods of working out how East and West can compete without war will be difficult enough.

For the strategic fundamentalists just identified, deterrence is more than simply a strategy: it is an ideology. As such, it shares the functions of other ideologies. Equally, in its aggravated form, it exhibits familiar negative symptoms. Let us examine these functions and symptoms.

THE FUNCTIONS OF DETERRENCE IDEOLOGY

Missionary Function

According to its followers, the most important issue in contemporary strategy is the theory and practice of nuclear deterrence. This belief is nurtured by an extensive body of doctrinal literature. Alternative approaches to achieving peace and security are shunned: people who do not accept the nuclear "theology" are seen as strategic primitives or heretics. Nuclear deterrence, after all, has a utopian character: its supporters believe that it will indefinitely ensure peace between the major powers. This dream is reinforced by reiteration of the belief that nuclear deterrence has already kept the peace for forty years. The enemy has been kept at bay. Nuclear deterrence has worked, is working, and will work.

Analytical Function

Once one's mind is infused with deterrence theory, everything falls into place. The world is explained by its circular and self-confirming logic: (1) we have faced a massive Soviet threat since 1945; (2) we have confronted them with a massive deterrent threat; (3) the Soviet Union has not expanded beyond the postwar settlement lines; (4) therefore, deterrence has been massively successful. What matters is the rigor of the theory, not a close analysis of the evolution of international history. Deterrence ideology consolidates a particular view of the world, and this, not its inherent truth or reasonableness, is what matters. Ideas about the rational conduct of nuclear war, with the object of achieving victory, can only be understood in ideological terms. Deterrence advocates see the world as they do partly as a consequence of the demands of theory. They need a deadly enemy, and thus the rigidities of bipolar thinking are reinforced. Consequently, the "keepers of the threat" in both superpowers have been more comfortable with the demonic, ideologically derived image of the enemy than with the more complex one offered by reality. For idealogues, it does not much matter what the enemy is like: it is only imperative to know that one exists. And absolute weapons demand absolute enemies.

Legitimizing Function

Deterrence has positive connotations. Not surprisingly, this has led it to become a slogan used in support of almost every doctrinal or technological twist or turn in the arms race. As long as it is done in the name of deterrence, almost anything can be justified, including the technology of overkill, war-fighting strategies that may make nuclear war more thinkable, and the illusion of prevailing in unwinnable wars. If it can be said to serve deterrence, it must be legitimate.

Operational Function

The theory of nuclear deterrence offers some guide to action, though in practice it has proved somewhat variable. As originally developed by Bernard Brodie and others, deterrence was associated simply with the idea of restraining direct attacks on one's own territory, or that of one's allies, by threatening to push up the costs of any aggression to an unacceptable level. That is no longer enough for many contemporary deterrence advocates. More utility is required for nuclear weapons, and the ultimate expression of this viewpoint is seen in the prescriptions of the supporters of counterforce strategies. The prevention of war is the immediate aim: but after that, nuclear weapons are to be employed for escalation dominance and intimidation short of war or, if war breaks out, for victory or at least to prevail. The addictive character of nuclear weapons, mentioned earlier, has been evident in the susceptibility of many deterrers to nuclear hypochondria—the mistaken belief that something is wrong, even if it is not. Such obsessions have displayed themselves in some parts of the United States defense community in the belief that the only important matter in the superpower relationship is the state of nuclear balance.[11] This has resulted in an unhealthy concern over the details of nuclear inventories, which, in turn, has been aggravated by a typically American obsession with statistics and notions of equilibrium. The outcome has been an unreasonable fear concerning what Albert Wohlstetter, in the late 1950s, called the "delicate balance of terror." This attitude was evident in the 1970s in the exaggerated fear that the Soviet Union would exploit the window of vulnerability that some Americans persuaded themselves had opened up in their deterrent posture. Deterrence addicts have an unbalanced terror of delicacy.

Socializing Function

Ideologies not only offer explanations and goals, but also provide language and symbols that bind their adherents. Deterrence theorists constitute a distinctive ideological (and sociological) group, with a strong ingroup–outgroup mentality. And as sometimes happens in such cases, an element of "groupthink" has developed among the deterrence community.[12] The latter problem is not one of intellectual shortcomings, but of group dynamics. Indeed, groupthink is associated with highly intelligent men with specialized knowledge who are both hard working and dedicated. In the case of the defense community there has been a tendency for its members to become oversocialized in the particular cluster of beliefs surrounding nuclear deterrence. They have their "mindguards" who protect the group's beliefs about the effectiveness and morality of their decisions. Alternative approaches are not taken seriously. The experts are defined narrowly, and dissidents are excluded from the real debate. Among those largely kept out by the intellectual circle of wagons drawn around deterrence theory have been Soviet specialists. They were not present at the creation of nuclear deterrence theory, and few have had any part to play in its dissemination. Deterrence theory has largely evolved in a geographical and historical vacuum.

Public Relations Function

Who can be against deterring nuclear war? Nobody. Consequently, deterrence has become a cloak behind which governments pursue an arms competition whose dynamic constantly undermines any concerted attempt to coexist. The label *nuclear deterrence* is used to sell weapons and policies to the general public that sometimes go beyond the strict needs of deterrence. As long as it is done in the name of deterrence, it has a seductive ring. A push for superiority, or merely a military edge, can be sought in the name of deterrence theory. These days, it seems, it is possible to advertise any posture under the label of *deterrence*. This is extremely welcome to those with a vested interest in big defense and those (an overlapping group) who see international politics in the crudest power terms.

Limiting Function

Ideologies establish boundaries of thought and action, and deterrence ideology has been particularly potent in this respect. Deterrence theory offers a static model of the world. Compulsive deterrers are locked within a particular set of theories and attitudes: their minds are constrained by the simple but powerful (self-validating) logic (or illogic) of deterrence. Their ideas about managing the superpower relationship are limited by the deterrent framework. They do not speculate outside it, but instead weave a dense pattern of ideas within it. Repetition is inevitable, and so is triviality. Deterrence advocates can be extremely inbred intellectually; they can cut themselves off from important dimensions of both the problem and possible ways of managing it. In this way, since strategies are supposed to be problem solving, the ideology of deterrence is irrational. Few deterrers, for example, seem to have discovered area studies, and fewer still have any particular expertise on Soviet matters, the ostensible target of all their effort. Nuclear deterrence is both a narrowing and negative preoccupation. Security depends on reassurance as well as deterrence, and the role of the former is often ignored: indeed, it is rejected by the strategic fundamentalists. It is sad to see some exponents of hard-nosed confrontation with the Soviet Union praised for the rigor of their narrow and negative thoughts. It is not difficult to be a dogmatic realist and to gain the reputation for being a clear and tough thinker. It is easier to be negative and confrontational than positive and constructive: it is easier to be clever about deterrence than wise about coexistence.

Consoling Function

Nuclear deterrence is a simple idea in an overcomplex world, and it promises peace in a world in which the cost of war is unthinkable to ordinary imaginations. Nuclear deterrence seems to solve all the problems in international politics in general, and in managing relations with the Soviet Union in particular. Nuclear deterrence simplifies: it takes the politics out of international politics and replaces it with mere technique. Furthermore, it bears almost the whole burden of the superpower relationship and so saves one from further worry about (or working at) the political problem of how West and East can live together without war for the indefinite future. The doctrine therefore provides intellectual, psychological, and practical comfort for those who give themselves up to it.

THE SYMPTOMS OF AGGRAVATION

Nuclear Deterrence Has Perpetuated Cold-War Thinking

Deterrence has helped institutionalize the cold war. Without minimizing the real clash of interests that exists between the superpowers, it is apparent that nuclear deterrence has encouraged threat inflation, rigid blocs, stereotyping, and overmilitarized interpretations of Soviet–American relationships. As long as both sides target each other with the technology of mass destruction and maintain the appropriate military–industrial establishments and rhetoric to sustain such postures, there will be no escaping the periodic chill of cold-war confrontation. Nuclear deterrence bolsters prejudice and fear and freezes old images. Despite the much more variegated picture of the Soviet Union now available, the strategic community as a whole is still shaped by the primitive Sovietology of thirty years ago. The clash of interests and suspicions that produced the superpower cold war have been magnified and perpetuated by the needs and effects of nuclear deterrence. This argument is given credence by the contrast between superpower relations and intra-European relations since the late 1970s. Detente died in the former case but not in the latter. The difference can partly (though not wholly) be explained by the pressures of nuclear strategy. This is not to say that the superpowers would live in harmony in the absence of nuclear weapons, were that possible (which it is not); it is simply to state that the preoccupation with nuclear technology has helped convert a normal clash of interests between the major powers of the day into a paranoid one.

Nuclear Deterrence Has Induced Complacency

Nuclear deterrence grew naturally out of the fears, interests, and technologies of the cold war, but it became overblown and is now irrational. The preoccupation with its details and intricacies obstructs the sensible formulation of the problems we face in world politics, including the handling of the superpower relationship. However, the apparent success of deterrence deflects criticism, as does the fact that it can be presented as being relatively cheap. Nuclear weapons, without doubt, represent powerful forces on the side of caution. Nuclear deterrence will, therefore, continue to work—until the very moment when everything goes wrong. At that point, the "elegance" of nuclear strategy will be of no help: it will also be too late to say no. Those who would protest must do so in advance. The lulling of critical faculties is reminiscent of the Ten Year Rule in Britain in the 1920s and early 1930s. Today, the West has a similar rule, except that it is implicit and its time span extends to a point in the future too distant to worry about. This assumption that there will be no general nuclear war releases one's sense of urgency, obstructs rational thought about security, induces complacency about the real problem, and discourages facing up to the imperatives of our time.

Deterrence Has Distorted Moral Values

Ideologies require an unquestioned belief in their own morality. This inhibits speculation about the ethical dimension of positions taken in support of the goals of the ideology. This was evident with most supporters of deterrence until the antinuclear movement forced at least some of them to articulate their position. It is right that moral issues—though not moralizing—form part of the strategic debate. Without moral lines drawn across the political confusion of contemporary strategy, we will have no standards of behavior and in no time, we will have reached the point toward which some are rapidly heading—namely, the socializing of nuclear war into our norms of international relations. This is the direction being followed by people who advocate deterrence by the threat of war fighting. Such an approach is based on anachronistic ideas, but is buttressed by neo-Clausewitzian rationalizations.[13] These war fighters insist on the relativity of catastrophe and on the moral imperative of nuclear targeting: their outlook has been labeled "counterforce morality."[14] It is sheer strategic casuistry. When many millions might be killed in a "limited" counterforce strike, and when such a strike would be one large step toward the end of civilized life in at least the northern hemisphere, the issue of nuclear targeting is not the key moral question that counterforce moralists insist it is. The key question is nuclear war itself. The important distinction is not between counterforce or countervalue (or various combinations or refinements on the basic targeting menu), but between war and peace with the Soviet Union. In this respect, counterforce strategies threaten to weaken rather than reinforce the prospects for peace. The road to hell may well be paved with good counterforce intentions.

Nuclear Deterrence Has Debased Strategic Language

Ideologies can twist language; words can be made to mean whatever their users wish. This has happened as a result of the nuclear addiction. The strategy of nuclear deterrence has been deprived of much of its original clarity. Deterrence is now a catch-all term that can mean or justify almost anything. What cannot be justified in the name of deterrence? We have: demonstrative deterrence (i.e., deterrence must not only be seen to be done, it must be done); intrawar deterrence (i.e., deterrence after deterrence has failed); and the threat of preemptive deterrence and the option of first nuclear use (i.e., deterrence before the fact of nuclear attack). If it serves deterrence, it is right: but what cannot be said to serve deterrence? This problem has been illustrated by Edward Thompson's criticism of the way nuclear deterrence has "spawned itself" as a noun. To speak of "the deterrent," in Thompson's words, is a "pathetic fallacy," preempting enquiry by attributing normative qualities to a weapon in the very same instant that the weapon itself is indicated.[15] A particularly blatant illustration of such usage has been the artless way President Reagan has attempted to label the MX missile as the "Peacekeeper." Who could object to strengthening the deterrent or to acquiring Peacekeepers? And who could object to increasing the numbers of such systems not only to enhance deterrence but also, so it is argued, to improve the prospects for arms control and thus to slow down the arms race. Clearly, we live in a strategic Wonderland, when the strategy of nuclear deterrence can be used to excuse, justify, cover, hide, rationalize, or sell almost any weapon or strategy, any intention or action. And morality follows the flag. Counterforce moralists in the strategic community and contextual moralists in the church have searched for a doctrine of the "just deterrent," to match that of the traditional doctrine of the "just war."[16] Like the just-war doctrine,[17] it seems that a just deterrent can justify anything.

NUCLEAR DETERRENCE VERSUS STABLE PEACE

History will record that nuclear deterrence began as a well-defined strategic concept, evolved into an ideology, and then, through addition, became a cult. As its grip grew on people's minds—and greater resources were committed in its name, and greater consequences were threatened—its value as a guide to action diminished. It is no longer the coherent and timeless strategic concept it was believed to be only a few years ago.

The criticisms of nuclear deterrence referred to earlier are not meant to suggest that the technology of mass destruction does not deter. It does, but it cannot be taken as fact that nuclear deterrence has worked in quite the manner that is usually asserted. All that can be said is that although nuclear weapons have undoubtedly imposed caution on the minds of decision makers, we do not know if they have been the only or even the main reason there has been no war between the superpowers since 1945. Any all-out war between industrial giants in the era of strategic bombing—even had nuclear weapons not been invented—would surely have threatened unacceptable damage to the participants. Nuclear deterrence has obviously not broken down, but we cannot say, and possibly will never be able to say, that it alone has prevented war. Even to the extent that it has worked in buttressing order, one might also ask whether it could have worked more cheaply, less dangerously, and with fewer long-term consequences in terms of aggravating superpower relations. If the aim of deterrence, like war, should be the creation of a "better peace,"[18] then nuclear deterrence as it has come to be practiced in recent years can hardly be judged a resounding success.

Unless one defines peace in the most negative sense—as the mere absence of war—then deterrence has not worked. It has not led to peace between the blocs. It has undoubtedly contributed to international order, however, and that in itself is important; but it is not the same as peace. Indeed, nuclear deterrence as practiced between the superpowers—involving a mutual threat to wipe each other off the face of the twentieth century—is antithetical to peace conceived in a positive sense. One has only to look at the state of Soviet–American relations forty years after the defeat of Hitler to appreciate that. Conflicts of interest and of perception exist in plenty, but nuclear deterrence and its baggage (arms competition, the pressures of weapons innovation, an overmilitarized interpretation of the confrontation, targeting strategies that threaten national survival, constant readiness, threat inflation, and so on) have played an important part in exciting high levels of mutual hostility and in poisoning the prospects for better relations through successive decades.

Deterrence ideology continually postpones the prospects for a stable peace, that is, one in which serious preparation for war does not take place because there is no serious expectation of war. We should have reached that position by now. There is no conflict of interest between the superpowers that would justify nuclear war. Since such a war would be an irrationality—the ultimate maladjustment of strategy and policy—it ought to be possible to rule out nuclear conflict between the superpowers as an instrument of policy. Mutual nuclear destruction by the superpowers would be a political absurdity: afterward, historians would wonder what causes could have justified such an outcome, and they would surely find them trivial. it would be a military nightmare: again, historians would wonder how such chaos and indiscriminate massacre could have been conceived as rational strategy. It would be a monstrous outrage against civilized life: historians would surely scorn those who could have claimed that the just deterrent was either legal or moral. It might also be an ecological catastrophe: those who lived in the aftermath would surely not forgive us. These are the parameters of our predicament. And the longer the opium of nuclear deterrence numbs our sense and distracts attention away from efforts seeking to create a stable peace between the superpowers, the greater will be the opportunities for Murphy, the malevolent god of strategy, to throw together that fateful combination of events and pressures that will make up a "nuclear Sarajevo."[19]

Changes are necessary on the part of both superpowers: but we have to begin where we are, in the West. Some, looking at Star Wars rising out of the far western sky, would say that profound changes are being made. They are mistaken. The so-called Strategic Defense Initiative (SDI) is neither inherently defensive nor ever likely to abolish nuclear weapons or ballistic missiles: such assertions are either fantasy or public relations. In practice, if Star Wars technology works passably well, it will merely be a fabulously expensive component in another and possibly less stable phase of nuclear deterrence. Unless both sides acquire equal confidence in the leakproof quality of their defensive umbrellas—an unlikely outcome—then Star Wars technology will provoke fears of intimidation, first strike ambitions, or preemption. With Star Wars, if it is ever deployed (and deployment can be stopped, even if research cannot be), the arms race will take another dangerous and depressing leap forward. Much time, energy, and money will have been invested in another failed technological fix to the problem of United States–Soviet relations.

The theory and practice of nuclear deterrence has been undergoing more than an artificial crisis in recent years. Never before has it been so beleaguered, being attacked by people on the political right (who now favor defense rather than deterrence) and people on the political left (who argue that deterrence has been superseded by war-fighting doctrines). Nuclear deterrence, both as an intellectual proposition and as a policy, has been in some disarray, and this has been both a cause and effect of the fact that the consensus on nuclear matters in several Western countries has become severely strained. Not only have there been great demonstrations in the streets, but a significant number of insiders and authorities have also become critics of established policy. Furthermore, there has been a growing sophistication within the antinuclear movements of Western countries; this has been evident in their attempt to transform the "ban the bomb" slogan into a valid set of alternative but realistic defense policies. Differences of outlook have also shown up within the strategic community itself, as has been seen in the Star Wars debates.

As a guide to action, what passed for deterrence theory has proved to be increasingly troublesome for many observers. In particular, growing numbers have been worried by the reasoning, the rhetoric, and the counterforce morality of the war fighting for deterrence advocates. With the advent of the Reagan administration, this school of thought achieved new political prominence, though the administration itself has so far adopted only a watered-down version of a true nuclear war-fighting strategy. Space forbids an extensive critique of the illusions of this approach. Suffice it to say that it places too much emphasis on both the technological dimension of security and the operational aspects of strategy; that it concentrates on the negative aspects of deterrence in the superpower relationship to the exclusion of almost all else; and that it exhibits a truly marvelous faith in the ability of what would remain of the United States political and military leadership to fight with nuclear weapons, control escalation, and bring the fighting to an end in a successful manner. But what if the plan fails? What if the enemy refuses to act, react, and collapse according to plan? And what if the *real* worst case materializes? In short, what has emerged from American think tanks as nuclear war fighting exhibits all the paradigm weaknesses shown by war planners through the ages. What makes the situation worse is the changed character of future war.

We should accept that nuclear war fighting is primarily advanced by its supporters to strengthen deterrence. Even so, by advocating that "Victory is Possible," that nuclear war should be fought in a "rational" fashion, and by suggesting that "catastrophe can come in different sizes," nuclear war-fighting theorists such as Colin Gray appear to have accommodated a strategy involving a political absurdity, a military nightmare, a moral outrage, and a possible ecological catastrophe into their norms of international politics.[20] The prescriptions of the strategic fundamentalists—albeit with deterrence in mind—increase the risks of war. Almost all observers now agree that a Soviet nuclear strike is highly unlikely as a deliberate act of aggression: anxiety is a more likely spur than aggressive ambition. This is where the prescriptions of the strategic fundamentalists have been most pernicious, for their words and policies have been calculated to increase Soviet edginess, fear, and suspicion. They have made an atmosphere of tension the norm of the superpower relationship. Unfortunately, deterrence theory, which copes reasonably well with the threat of deliberate aggression, is weaker when dealing with less rational behavior. Furthermore, those scenarios that have become political realities in recent years—the Lebanon bloodbath, the Iranian *jihad*, the Falklands escalation—have owed less to cool heads than to national or ideological passions, feelings about "face," gut reactions, and the valuing of justice over safety. When cool heads do not prevail, calculations about costs and benefits change, and deterrence theory becomes less elegant. The strategic fundamentalists in the United States have strained—and if they have their way will strain even more—the capacity of Soviet leaders to exercise sober judgement. They have rejected Bernard Brodie's sound advice about the undesirability of pursuing marginal advantages in fighting nuclear wars at the cost of undermining the stability of nuclear deterrence.

At the same time that the technology of mass destruction is being socialized into a Hobbesian view of the world and arguably making war more likely by encouraging us to prepare psychologically and practically to fight and prevail, the utopian rule of deterrence ideology is pushing in the other direction by offering consolation and by suppressing the reality of the risks of nuclear war. Conveniently, the dream of an effective Star Wars system in the hands of the United States serves both the operational and utopian requirements of nuclear strategy. It promises to meet the desire of the strategic fundamentalists to recover the utility of military power by offering real defense, while it satisfies the American nostalgia to recapture total security. Both groups look forward to more security based on American strength and technological prowess; and they have reduced their estimate of the risks of the worst case virtually to nil. Not for the first time in the twentieth century, our ostensible worst-case forecasters have closed their minds to the real worst case.

IMPENDING RISKS

No strategic doctrine can guarantee that it is free from the possibility of breakdown. Powerful though it is, nuclear deterrence is no exception. Indeed, the concept itself rests on the belief that in some circumstances the threat to engage in nuclear war will be carried out. Despite this, for much of the nuclear era, the theory and practice of Western strategy has paid relatively little attention to the actuality of nuclear technology's ever being used in anger. Metaphorically speaking, we have thrown up our hands in horror, thus serving to underline Robert Lifton's "rule of thumb" that "the more significant an event, the less likely it is to be studied."[21] However, this might have begun to change somewhat at the turn of the 1970s to the 1980s, when more people came to believe that the structure and processes of international politics were inclining the system in a more dangerous direction. Though certainly not everybody agreed, a substantial body of opinion came to recognize that we now have a less well-constructed superstructure of nuclear deterrence resting on foundations whose geology is less predictable than formerly. The risks of nuclear war are not overwhelming, and they can be reversed: but they will not be until more of us are willing to recognize that there are dangers that should be reversed. The risks we face, which constitute the argument for thinking that we are facing a difficult period ahead, are political–strategic dangers and military dangers.

Political–Strategic Dangers

Soviet Insecurity. If history is not seen to be going its way, will Soviet policy become more militant? Will it blunder into confrontation?

American Nationalism. In reaction to the humiliations of the recent past, will resurgent American nationalism tempt the United States into overstretching its power and taking excessive risks? Will it push its rival into a corner?

Conflict over Resources. In a period of hardship, is an attitude of "every man for himself" likely to lead to conflict over the world product?

Nuclear Proliferation. Will nuclear proliferation take place in regions of chronic instability? Can we expect regional nuclear wars? Might such wars draw in the superpowers?

The Breakdown of the Sense of International Community. Will the growth of political, economic, religious, and cultural strains among nations lead to a general rise in tension, and with it a rise in risk taking? Will harsh times encourage "crazy states"?

The Rise of Interventionism. Will the incapacity of many states result in a season of intervention? Will it always be possible to contain local crises?

Superpower Competitiveness. Will either the United States or the Soviet Union always be willing to stand by while the other intervenes? Might they one day become embroiled? Will one or both of them always find it possible to back down?

The Decline of Spheres of Influence. What are the chances of a collapse of superpower hegemony in Eastern Europe or Central America? Should these regions become spheres of impotence, would the adversary exploit the situation? Would a threatened collapse stimulate dangerous behavior?

The Decline of "Crisis Stability."[22] If the superpowers become embroiled in a crisis, can we be as confident as formerly—and forever—that their ability and determination to manage the crisis will prevail over their desire to prevail, their propensity to miscalculate, or their fatalistic sense that they have no choice but to press on up the ladder of escalation?

Military Dangers

Soviet Inferiority. If Soviet planners believe that by the 1990s the United States will have a growing potential to exercise escalation dominance, conduct limited nuclear options, execute a first strike, or prevail in war, will they be provoked into preemption?

The Shift to Intimidation in United States Nuclear Doctrine. With deterrence credibility in the United States becoming equated with the threat of the successful conduct of nuclear operations, must we face an indefinite future of increased arms competition and mutual suspicion? If the notion of sufficiency in the arms race is obsolete, does this rule out the prospect of security through arms control? Can there be security without mutual restraint in armaments policy?

The Erosion of the Nuclear Taboo. Does the talk of surviving and of prevailing[23] mean that the taboo[24] against nuclear war has disappeared? Instead of the first nuclear war being the result of accident, miscalculation, or madness, might it be a scientifically calculated disaster?

The Growth in the Expectation of Nuclear War. Will the widespread fear of nuclear war be self-fulfilling? If any decision maker comes to believe that Armageddon is inevitable, will it become so?

The Development of First Strike Strategies. Although a first strike out of the blue remains an unlikely possibility, might one or the other superpower persuade itself in a crisis that striking first will be preferable to retaliation? Given some of the technology available, will military planning that encourages haste encourage war?[25]

Adherence to a First Use Posture. Could a general nuclear war result from a combination of NATO's exaggerated perception of its conventional weakness and its operational and psychological commitment to the doctrine of first use? Will it therefore be the case, to adapt a line from Susan Sontag, of "*Cogito* first use *ergo* boom"?

The Risks of Accidental War. Might nuclear war occur as a result of the malfunction of critical technology? Is there a serious risk of multiple interacting errors, especially in a period of tension, with reinforcing superpower alerts encouraging both sides to interpret the actions of the other as evidence of an impending attack?[26] Might the superpowers become prisoners of their technology and standard operating procedures? Will the nuclear timetable of today set in train the race from a nuclear Sarajevo to a nuclear Armageddon?

The Illusion of Limited Nuclear War. Does planning for nuclear options make nuclear war more likely? Does it give an entirely false sense of reasonableness and manageability to what would end up as a lunatic nightmare? Is the idea of controlled nuclear war the most dangerous illusion of all?

Unstable Battlefield Doctrines. Does the forward deployment of battlefield nuclear weapons increase the likelihood of escalation? Will the pressure of "use it or lose it" preempt decision makers in the very early stages of war?

Offensive Developments in Conventional Doctrine. Are recent developments (such as Star Wars and emerging technology) making Europe both safer for conventional war and at the same time making nuclear war more likely by blurring the nuclear–conventional threshold? Will the traditional Soviet emphasis on offensive operations, together with recent changes in NATO's conventional doctrines, encourage a foolish overconfidence that may one day provoke nuclear retaliation?

Technical Breakthroughs. Will a technical breakthrough provoke the weaker side to preempt or tempt the stronger side to engage in preventive war? Does the possibility of Ballistic Missile Defense threaten the very stability of deterrence? Will the Star Wars initiative provoke the "Evil Empire" to strike back (or first)?

The Dangers of the Arms Race. By encouraging the production of destabilizing weapons, overmilitarized interpretations of the superpower relationship, and mutual threat inflation, is not the arms race the main cause of a cycle of mistrust that is a growing threat to peace? Although arms races do not inevitably lead to war, is it not true that wars have often been preceded by arms races? Will the tension of the contemporary arms race become unbearable? Will the strain cause somebody to crack?

One does not have to be an alarmist to recognize the relevance of such questions about the international situation. Of course, one need not give the pessimistic answer in each case: but one has only to concede that these have been and remain serious questions in the

1980s (they were barely relevant for the first half of the 1970s) to appreciate the significance of the change in our level of insecurity. Nuclear weapons remain impressive deterrents, but the questions just posed scorn the complacency of those who believe that we have finally acquired a deterrent that will always work.

Our discussion of these matters is greatly hindered by the fact that we do not know how to calculate the risks of any war, let alone nuclear war. Risk analysis, like threat assessment, is one of the most primitive areas of statecraft. In any case, we cannot forecast the likelihood of nuclear war statistically, for we have no observed frequencies on which to base our calculation. We can only back our guesses with numbers.[27]

When it comes to guesses, authoritative voices have been heard on both sides of the debate. Lord Mountbatten's warning about the nuclear "precipice" is now famous, as is the further movement toward midnight of the doomsday clock on the cover of the *Bulletin of the Atomic Scientists*. On the other side of the debate, it is just as strongly argued that the risk of nuclear war has not increased, that there may never be a nuclear Sarajevo, that arms races do not necessarily lead to war, that nuclear deterrence threats are powerful, and that technical breakthroughs are unlikely. One apparently comforting assumption arises out of the difficulties people face in conceiving doomsday scenarios and initiating nuclear strikes in war games. There is some consolation here, but not much when one thinks of the catastrophic consequences of only one error. Furthermore, war games are only *games*; they are not real life. People behave relatively coolly, and their stake in the outcome is small. In real life, one's emotions are engaged, and calculations of costs and benefits change. Even the most cleverly constructed war games can tell us only a small amount about the way we might feel and calculate under the pressure of an intense crisis.

Our inability to conceive entirely plausible war scenarios is nothing new. Nobody guessed the train of events that would follow from Archduke Ferdinand's visit to Sarajevo in 1914. The resulting war was a surprise—the international outlook in Europe in the summer of that year had been far from thundery—but it was scarcely an accident. The war erupted out of the reaction between the structure and processes of the contemporary international situation. The decision makers of the time were exposed, in a period of tension, as being the playthings of ambitions, emotions, misperceptions, miscalculations, plans, and the dynamics of events. There is no reason to suppose that we are any wiser, or that we and those who follow will be forever luckier than our forefathers. When, against a background of international tension, fear is mixed with a desire to show commitment, and when the advantage of striking first is combined with faith in success, the interplay of men and machines can be fatal. We cannot predict the doomsday scenario, but we can forecast that, like the poor, international crises will always be with us. And it only needs one to go out of control. Oddly, it is the self-professed realists, those who most pride themselves on their understanding of the gloomy path of human history, who seem unable and unwilling to believe that the deterrent may one day fail to deter and that the balance of nuclear power may one day collapse.

DISSUASION *A OUTRANCE*

Clausewitz, according to some "the apostle of violence," would probably have found himself unemployed in the world of contemporary strategy because of his squeamishness. Like Sun Tzu and Thucydides, he would have recognized the business of politics among nations—the search for power, the formation and disintegration of alliances, and so on—and also like them, his mind would have been blown by the extreme destructiveness of devices now considered to be weapons of war. Yet it is those very devices the strategic fundamentalists are trying to adapt to what they think are the needs of the Clausewitzian philosophy of war. One of the greatest ironies of our predicament is that nuclear strategists who have tried to face up to the realities of the strategic predicament have aggravated the problem rather than found a cure. In the nuclear age, it is doubtful whether Clausewitz himself would have remained Clausewitzian.

We should guard against being lulled by the Clausewitzian overtones of the war-fighting school into overlooking the scale of their ideas: they are planning, as a rational act of policy, not only to threaten but also in some circumstances to engage in a mutual exchange of violence unprecedented in world history. At this point, whatever the war fighters claim, we are no longer talking about strategy. Such an outcome cannot be considered war in a traditional sense. It would be a nihilistic act of violence pushed to its utmost bounds; and it is difficult to conceive how such an extreme of violence could be considered a continuation of politics. What price a Clausewitzian inspiration, therefore, if the result is an ability to think strategically, but at the cost of an ability to think politically, let alone in humanitarian terms? War has always corrupted, but absolute war corrupts absolutely. What today's strategic fundamentalists recommend emerges from national biases and ideological beliefs rather than from objective and empirical analysis. For them, strategic reasoning is not only a matter of adjusting ends and means but also a cloak for an ethnocentric world view. The possibility that they do not know what they are doing is just as worrying as the chance that they do.

When warfare was relatively limited, it was reasonable to believe that strategy was exclusively a matter for professionals. The Great War changed that, and nuclear weapons have taken us a stage further by universalizing violence. Nuclear weapons democratize: we all become equal in victimhood.[28] Strategy less than ever is simply a matter of considering the interplay of weapons and doctrines. Today, it is vitally concerned with basic questions about the nature of war, the meaning of security, and the role of the national interest. Strategy is a continuation of moral philosophy with an admixture of firepower.

More than ever, we need to pay attention to Ferdinand Foch's little question (made prominent by Brodie): "*De quoi s'agit-il?*" What is it all about?[29] The fundamental point in this respect is that before any act of violence can be considered a continuation of politics, it must first give some guarantee that its aftermath will allow a continuation of civilized life. Here, the women at Greenham Common, so derided by the defense establishment in Britain, have touched a sensitive nerve. Even if their military expertise is limited, their contribution to the defense debate has been a vital corrective to the outlook of the many strategically illiterate party hacks and the closed world of many nuclear deterrers. When the Greenham women pinned homely trinkets or children's photographs on the perimeter fence of the cruise missile base, they were trying, in their way, to answer Foch's question: they know that the language of the nuclear strategist is not enough. Einstein also knew it. When asked what would happen in the event of nuclear war, he ignored the obvious temptation to talk about it in the currency of strategy and instead tried to point to its deeper meaning: "Alas," he is supposed to have said, "we will no longer be able to listen to the music of Mozart."

In the past, bearers of bad news were sometimes killed. In the nuclear debate in the 1980s (as has happened with the Campaign for Nuclear Disarmament in Britain or Ground Zero in the United States, for example), those who have challenged the verities of nuclear deterrence have been derided for being gloom and doom merchants, zapped with official propaganda, or defined out of the real debate. This is the price that sometimes must be paid for trying to question established beliefs. It would be a mistake, of course, to assume that the dissidents always get it right in strategy; that would be as misguided as believing that the government always knows best. The peace movement, for example, got it wrong in the 1930s, but so did almost everybody else. The implication of these points is that we need a more constructive dialogue between the various sectors in the defense debate: between realists and idealists, short-term policy makers and long-term planners, unilateralists and multilateralists, technicians and philosophers. Strategic history is littered with the casualties resulting from a failure to share ideas. Traditional strategic reasoning alone cannot be relied on to get us out of our predicament, for it has played a large part in getting us where we are.

Today, newsreels of only forty-five or seventy years ago look incredibly old-fashioned. We stare in fascinated horror at scenes in which people we know may have participated. If NATO's doctrine of flexible response (i.e., inflexible escalation) were ever put to the test, how would it appear on the newsreels of the next few generations? What would our grandchildren think about a plan involving the sending of men against nuclear fire? They would surely be both fascinated and horror-struck. Yet we are not evil, nor particularly stupid. We are simply in a deep trap and no wiser then our strategic forefathers. Rumors of war and echoes of the past are all around us. In peacetime, the ghost of Archduke Ferdinand stands in the wings; the memory of the Great War shows the difficulty of disengagement once emotions become heated, underlines the fact that men are not deterred by fear, and remains an object lesson in the dangers of sticking to ideas for too long. So, before it is too late for us, we must do what we can to nudge strategic history out of the grip of an ideology of deterrence, which threatens to socialize nuclear war, and away from a philosophy of war that technology has made obsolete.

In view of the dangers we face, and the difficulties of getting out of them, it is evident that the agenda of strategic studies will remain full. There is no fear of redundancy for strategists when we see: (1) the trend toward the socialization of nuclear war; (2) the gap between the technology available (of uncertain reliability) and the elegance (as some would have it) of strategic theory; (3) the ethnocentric incomprehension of adversaries from the superpowers downward; (4) the need to work toward a stable peace rather than one largely based on the threat of mutual massacre; and (5) the urgency of freeing humanity at large from the threat to their existence represented by the possible escalation of a quarrel between the governments of only 12 percent of the world's population.

When it comes to thinking about what can be done, we have to start from where we are. We do not have the option of changing the game of nations. Nor can we define the Soviet problem out of existence or remove nuclear weapons from the face of the earth. All we can do is try to agree, tacitly or formally, to play the game more safely. This means mixing realism and reform. Space forbids any detailed discussion of prescriptions, but in the light of the central theme of this chapter, that is, the grip of deterrence ideology and the finite possibility of a catastrophic breakdown, there are steps that can be taken. Deterrence, in some form or another, has always been with us and always will be, but in both its present theory and practice it needs demystifying, demoting, and denuclearizing.

The nations in the central strategic confrontation have no sensible alternative but to work at what might be called reciprocal security.[30] Because the superpowers have the destructive power to wipe each other out, and much else besides, and because neither has a guaranteed defensive capability in sight (Star Wars technology immediately begat work on anti–Star Wars technology), traditional ideas about national security based on relative national strength will not suffice. Safety can come only through relationships, and secure relationships involve reciprocity. Among nations, reciprocity implies the mutual exchange of privileges; and in the late twentieth century, the greatest privilege nations can exchange is that of security—the fundamental right to exist, independently and free of the risk of genocide or domination.

The achievement of such a goal should place the denuclearization of United States–Soviet relations at the top of the agenda of security problems. This does not mean trying to eradicate nuclear weapons. Such an aim is a snare and a delusion, a distraction from what can be achieved, namely, reducing the prominence of nuclear weapons in the superpower relationship. For the immediate future, the priority should be to try to move the character, deployment, and doctrine of nuclear weapons in a less threatening direction: but even this is easier said than done. The conflicts of interest between the West and the Soviet Union cannot be wished away, nor can we ignore the Soviet intention of competing vigorously with the West. But we should never underestimate Soviet rationality. It is thirty years, after all, since Nikita Khrushchev revised the Leninist prophecy about the "inevitable" clash between the systems. Today, fatalistic talk about Armageddon comes from the right wing of American politics.

There are grounds for hoping that reciprocity can work in the security field: both superpowers, and their allies, have objective interests in ensuring that it does. For a start, this means reformulating the central problem of superpower relations from What deters? to How do we live together? With this in mind, a variety of reforms have been placed on the agenda. These are based on an acceptance of the inescapable reality of the situation of mutual assured destruction between the superpowers (whatever the prevailing doctrine or Star Wars rhetoric) and of the need for nuclear restraint. Among the reforms that have been put forward to denuclearize the superpower relationship, the following have been the most prominent:

1. A comprehensive nuclear test ban.
2. The adoption of a no-first-use nuclear posture (or at least a no-early-first-use as a preliminary step).
3. A superpower nuclear freeze.
4. Increased efforts to inhibit nuclear proliferation.
5. Unilateral arms control initiatives.
6. Prevention of the growth of weapons competition in space.
7. Exercise of unilateral and if possible multilateral restraints in technological innovation (especially with destabilizing systems such as MIRVs).
8. The strengthening of conventional deterrence options to raise the nuclear threshold.
9. The progressive withdrawal of battlefield nuclear weapons from the forward zones on the Central Front.
10. The pursuit of confidence-building measures.
11. The creation of both political and military fail-safe devices (crisis control centers and effective command and control systems).
12. Movement away from strategic doctrines involving preemption, launch-on-warning, decapitation strikes, and victory, toward doctrines in which retaliation is a last not a first step, gradual rather than massive, and based on what is considered sufficient for a minimum deterrent rather than what is thought necessary for war fighting.

In addition to exploring such reforms, relations with the Soviet Union need to be redirected from the overmilitarized confrontation of recent years to a relationship that is wider and more positive. This means less confrontation and more cooperation, less emphasis on deterrence and more on foreign policy, and less military rivalry and more exploration of agreed norms of behavior. To further these aims, important voices in the United States should reject the old-style Sovietology of the new right and recognize the changes that have taken place; and those in power must change the rhetoric in their relationship with the Soviet Union and treat it as a superpower equal. At the same time, trade and cultural relations should be increased to the greatest possible extent (not in the hope that they will guarantee stability but as tokens of reassurance). By undertaking such actions, a more stable political relationship should become established in which crisis prevention, not cold war, is the norm.

It will not be easy to persuade the governments of the Western powers, let alone the Soviet Union, to accept the desirability of some or all of these reforms. Powerful forces in both camps find some degree of confrontation to their taste. Hence, the pursuit of reciprocal security will not be easy nor without failure. Some reforms may prove less useful than others; but one thing is certain: simply throwing more megatons at the problem of Soviet–American relations will not help.

If we do not try to change the character of the nuclear confrontation, history will surely judge us badly, whatever the outcome. Robert Oppenheimer once said that the nuclear age would be seen as a phase in human history that was transitory, dangerous, and degrading. Either it would be understood in this way, or there would be no one there to make the judgement.[31] If one rejects the criticisms that have been made of our nuclear postures, think how our strategy would look from the viewpoint of a historian after World War III. Or, if we were to be on the losing side in that event, how would we defend our early and heavy commitment to nuclear strikes before a latter-day Nuremberg war crimes trial?[32] What would be our defense? Even if the deterrent does continue to deter forever more and the worst case is avoided, future historians will no doubt look back and think of the lost opportunities, the risks, the wasted skills, and squandered resources while so much else needed to be done. If those historians happen to be Western, they might be shocked by the evil we threatened and were prepared to carry out in the name of defending civilized values. So what? These are the burdens the nuclear deterrers tell us we must accept, because the world is a dangerous place and we must prevent Soviet domination, either through military conquest or blackmail. However, many of us—including most who study the Soviet Union—believe that the Soviet threat and the dangers of instability have been greatly exaggerated. Consequently, rather than pursuing security through accretions of national strength, which only bind us tighter in the trap, the risks can be better managed by security policies in which nuclear weapons have been drastically reduced in prominence. Rather than wasting yet more time and effort in creating further obstacles to our escape from the war trap with our adversary partner, we should be using the margin of safety, which presently exists as a result of the deterrent effects of nuclear weapons, in order to explore the potential for living together with greater reciprocal security.

History will not judge deterrence ideology well. It will see it in much the same way that we look on the French "theory of the offensive" before World War I. Future historians, like those who have studied the French cult of the offensive, will struggle to understand the historical, psychological, bureaucratic, and professional reasons a traditional military idea and disposition were pushed too far. They will shake their heads at the excessive optimism displayed by nuclear deterrers; they will criticize the foolish bravery involved in planning to pitch men forward on nuclear battlefields; they will wonder about the decades of incuriosity shown by Western strategists toward their superpower enemy; they will be amazed that talk about victory could be taken seriously in the nuclear context; they will shake their heads over the assumption that all will go according to plan; they will puzzle over how people could so distort politico–strategic reality in order to nurture their beliefs and biases; and they will note that, not for the first time, "bad anthropology" made "bad strategy."[33] If nuclear war does occur, those future historians with the desire to reason will ponder the crazy courage, the extent of the clash of interests that could have provoked such an outcome, the disparity between costs and results, the fact that balances of power do not always produce peace, the gap between strategic theory and technological reality, the desirability of an early draw over a pyrrhic victory, the reminder that deterrents do not always deter, the fact that some doctrines can encourage what they are intended to prevent, and the recurring habit of worst-case forecasters to ignore the real worst case. In short, future historians will see close parallels between the cult of the offensive and the cult of nuclear deterrence, between *l'offensive à outrance* and today's *dissuasion à outrance*.

Bernard Brodie described the cult of the offensive as "a case study in military aberration," but a kind of aberration that in one form or another is always reappearing.[34] It was a case, he said, of an idea that had a "sound basis in reason," but that by "constant and fervent reiteration" had created distortions. The same is true of the cult of nuclear deterrence. It had a "sound basis in reason" when it began, but subsequently it became an addiction. The supporters of *l'offensive à outrance* hoped to be Napoleons in the age of efficient machine guns and artillery. The disciples of *dissuasion à outrance* attempt to be Clausewitzians in an era that is overburdened with the technology of mass destruction. War between the major industrialized powers, both functionally and semantically, is obsolete. Yet:

> They do not understand
> that war is dead and tomorrow
> will bring only a collision of deterrents.[35]

NOTES

1. Paul Fussell, *The Great War and Modern Memory* (Oxford: Oxford University Press, 1975), p. 74.

2. On the difference between a "doctrinal realist" and an "empirical realist" see William T. R. Fox, "E. H. Carr and political realism: vision and revision," *Review of International Studies*, Vol. 11, 1985, pp. 12–13:

> The doctrinal realist does not need to examine how nation–state actors really behave because he has already posited how they *must*

> behave by virtue of their nation-stateness in a multi-state world system. The empirical realist by contrast looks to see how they actually behave and perhaps beyond that to examining the possibilities for tolerable coexistence.

The distinction will be employed on several occasions in this essay.

3. Lewis Thomas, *Late Night Thoughts on Listening to Mahler's Ninth Symphony* (Toronto: Bantam Books, 1984), p. 6.

4. The most notable example is Sir John Hackett, *The Third World War: August 1985. A Future History* (London: Sidgwick & Jackson, 1978).

5. For summaries of the "golden age," see Ken Booth, "The Evolution of Strategic Thinking," in John Baylis et al., *Contemporary Strategy. Theories and Policies* (London: Croom Helm Ltd., 1975), pp. 22–49; Lawrence Freedman, *The Evolution of Nuclear Strategy* (London: Macmillan, 1981); Colin S. Gray, *Strategic Studies and Public Policy* (Lexington: The University of Kentucky Press, 1982).

6. Lawrence Freedman believes it is an "artificial crisis": see his "Indignation, influence, and strategic studies," *International Affairs*, Vol. 60 (2), Spring 1984, p. 217.

7. The literature is now very extensive. An excellent place to start is Patrick M. Morgan, *Deterrence. A Conceptual Analysis* (Beverly Hills, Calif.: Sage Publications, 1983).

8. E. P. Thompson, "Deterrence and Addiction," pp. 1–25 in *Zero Option* (London: Merlin Press, 1982); and Robert J. Lifton and Richard Falk, *Indefensible Weapons: The Political and Psychological Case against Nuclearism* (New York: Basic Books, 1982).

9. See Thompson's comments on the remarks of Professor Gregory Bateson: Thompson, "Deterrence and Addiction," pp. 2–3.

10. The journal *Commentary* has been an important public source of the writings of the new right. See, for example, Norman Podhoretz, "Making the World Safe for Communism," *Commentary*, April 1976; Richard Pipes, "Why the Soviet Union Thinks It Could Fight and Win a Nuclear War," *Commentary*, July 1977; Eugene V. Rostow, "The Case against SALT II," *Commentary*, April 1980. A revealing symposium, comprising about twenty of the best and rightest, was published in The Heritage Foundation's *Policy Review* (No. 31), Winter 1986: "Beyond Containment? The Future of U.S.–Soviet Relations." The most prolific strategic fundamentalists in the academic literature have been Colin S. Gray and Edward N. Luttwak. See inter alia, Gray's "Nuclear Strategy: The Case for a Theory of Victory," *International Security*, Vol. 1(4), Summer 1979, pp. 54–87; with Keith B. Payne, "Victory Is Possible," *Foreign Policy* (No. 39), Summer 1980, pp. 14–27; "War Fighting for Deterrence," *The Journal of Strategic Studies*, Vol. 7(1), March 1984, pp. 5–28; and *Nuclear Strategy and Strategic Planning* (Philadelphia: Foreign Policy Research Institute, 1984). Note William R. Van Cleave's Foreword to the last mentioned work. For samples of Luttwak's work, see inter alia, *The Grand Strategy of the Soviet Union* (London: Weidenfeld and Nicolson, 1983), and *Strategy and History. Collected Essays*, Vol. II (New Brunswick, N.J., and Oxford: Transaction Books, 1985). Intriguing insights into the mind-set of several

important members of this body of opinion are offered in Robert Scheer, *With Enough Shovels: Reagan, Bush & Nuclear War* (London: Secker & Warburg, 1982). The ideas of the strategic fundamentalists are not those of a fringe group of uninfluential people. They are at the heart of the Reagan administration's policy. This was evident, for example, in the Pentagon's 1984–88 Defense Guidance Plan, which was leaked to the press: Richard Halloran, "Pentagon Draws up First Strategy for Fighting a Long Nuclear War," *New York Times*, May 30, 1982; and Robert Scheer, "Pentagon Plan Specifies Methods of Winning Protracted Nuclear War," *International Herald Tribune*, August 16, 1982. For surveys of this plan see Christopher Paine, "Nuclear combat: the five-year defense plan," *The Bulletin of Atomic Scientists*, November 1982, pp. 5–12; and Phil Williams, "Deterrence, Warfighting, And American Nuclear Strategy," *ADIU Report*, Vol. 5 (No. 1), January/February 1983, pp. 1–5. Important members of the Committee on the Present Danger moved into the administration. To cite a few examples, Kenneth L. Adelman became Director of the Arms Control and Disarmament Agency (succeeding his fellow committee member Eugene V. Rostow), Colin S. Gray became a Member of the General Advisory Committee on Arms Control, Fred Charles Ikle became Under Secretary of Defense for Policy, Max M. Kempelman became Chairman to the U.S. Delegation to the Conference on Security and Cooperation in Europe, Richard N. Perle became Assistant Secretary of Defense for International Security Policy, Richard Pipes became a staff member of the National Security Council, etc., etc. Ronald Reagan became president, twice.

11. This point is well made throughout George F. Kennan, *The Nuclear Delusion: Soviet–American Relations in The Atomic Age* (London: Hamish Hamilton, 1984).

12. See Irving L. Janis, *Victims of Groupthink. A Psychological Study of Foreign Policy Decisions and Fiascos* (Boston: Houghton Mifflin Co., 1972).

13. The phrase was coined by Anatol Rapoport, *Clausewitz on War* (Harmondsworth: Penguin Books, 1968), pp. 54–80. Rapoport is attaching the label *Clausewitzian* to modes of thought that, in terms of this essay, are decidedly un-Clausewitzian.

14. Lifton and Falk, *Indefensible Weapons*, p. 38.

15. Thompson, "Deterrence and Addiction," p. 6.

16. Rev. Richard Harries, ed., *What Hope in an Armed World?* (London: Pickering & Inglis, 1982), p. 108; and his "In search of a just deterrent," *The Times*, March 1, 1984.

17. See Donald A. Wells, "How Much Can the 'Just War' Justify?" *The Journal of Philosophy*, Vol. 66, No. 23, December 1969, pp. 819–29.

18. See Michael Howard, ed., *Restraints on War* (Oxford: Oxford University Press, 1979), p. 14.

19. The phrase is Paul Bracken's, *The Command and Control of Nuclear Forces* (New Haven, Conn.: Yale University Press, 1983), p. 2.

20. For a sample of Gray's work see note 10. For a humane critique of this approach see Michael Howard, "On Fighting a Nuclear War," in his *The Causes of Wars* (London: Unwin Paperbacks, 1983), pp. 133–50.

21. Lifton and Falk, *Indefensible Weapons*, p. 38.

22. "Crisis stability" is another contradiction in terms, which some strategists find consoling.

23. See the revealing interviews in Scheer, *With Enough Shovels.*

24. "The taboo against nuclear weapons is the only protection we have." Carl-Friedrich von Weiszacker, "Can a Third World War Be Prevented?" *International Security*, vol. 5(1), Summer 1980, pp. 198–205.

25. See Thomas C. Schelling, *Arms and Influence* (New Haven, Conn.: Yale University Press, 1966), pp. 227–32.

26. Bracken, *The Command and Control*, pp. 2–3.

27. For example, Jim Penman, "What chance Armageddon?" *The Guardian*, October 10, 1983.

28. Freeman Dyson, *Weapons and Hope* (New York: Harper & Row, 1984), discusses the gap between the "world of the warriors" and the "world of the victims," pp. 4–9.

29. Bernard Brodie, *War and Politics* (London: Cassell, 1973), chapter 1. The primacy of objectives, of course, infused Clausewitz's work.

30. This is discussed further in Ken Booth, "The Case for Non-nuclear Defence," esp. pp. 52–5, in *The Future of British Defence Policy*, ed. John Roper (London: Gower Publishing Company, 1985).

31. Quoted by Gwyn Prins, ed., *The Choice: Nuclear Weapons Versus Security* (London: Chatto & Windus, 1984), p. xvii.

32. This idea is suggested by Dyson, *Weapons and Hope*, p. 163.

33. Bernard Brodie, *Strategy in the Missile Age* (Princeton, N.J.: Princeton University Press, 1959), p. 52.

34. Ibid., p. 43.

35. From Edwin Brock, "The Heroes," *Encounter*, April 1983, pp. 42–3.

Index

T

For Product Safety Concerns and Information please contact our EU representative GPSR@taylorandfrancis.com
Taylor & Francis Verlag GmbH, Kaufingerstraße 24, 80331 München, Germany

www.ingramcontent.com/pod-product-compliance
Ingram Content Group UK Ltd.
Pitfield, Milton Keynes, MK11 3LW, UK
UKHW022219190926
475702UK00003B/439
* 9 7 8 0 3 6 7 5 3 2 4 5 1 *